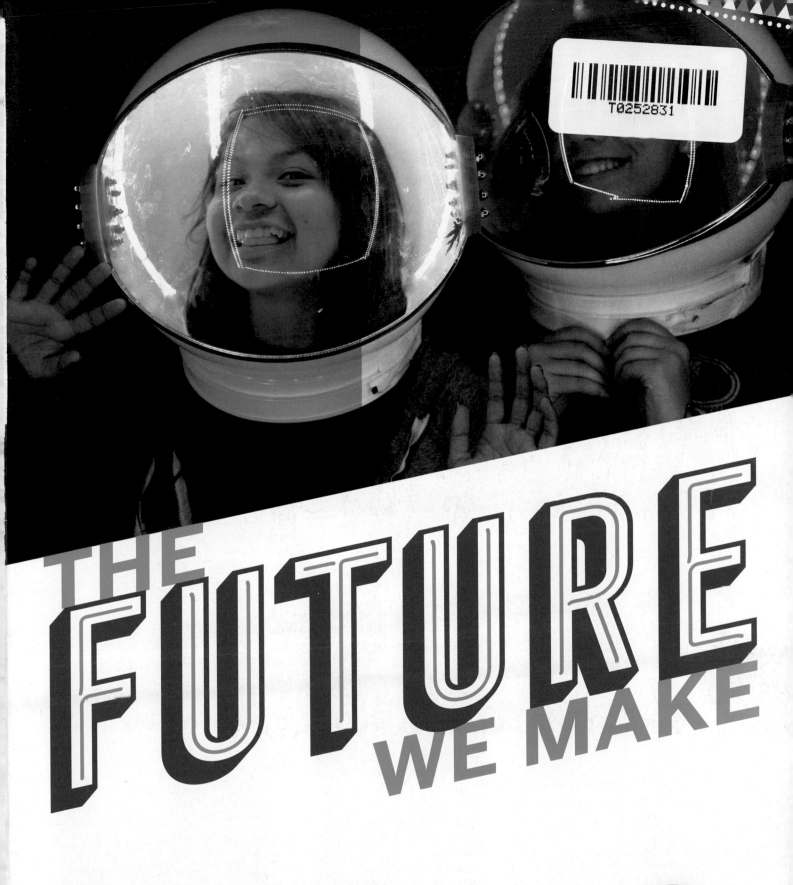

THE FUTURE WE MAKE

Maker Faire® MAY 17–19

BAY AREA SAN MATEO EVENT CENTER
makerfaire.com

T0252831

CONTENTS

Make: **Volume 69** June/July 2019

30

38

40

ON THE COVER:
NASA's Astrobee robots are ISS-bound semi-autonomous helpers that students can access and control on the space station, as part of the Zero Robotics competition.

Artwork: NASA and Yuriy Mazur-Adobe Stock (background)

43

Joe Barnard/BPS.space, Caleb Kraft, NASA, Ken Condal, NASA/Clouds A0/SEArch+, NASA/Cornell University, Kim Pimmel, Debra Ansell, Josh Lewis, Adobe Stock – kungverylucky, Guido Bonelli, Bob Knetzger

48

50

52

56

62

66

72

74

Make:

EXECUTIVE CHAIRMAN & CEO
Dale Dougherty
dale@makermedia.com

CFO & COO
Todd Sotkiewicz
todd@makermedia.com

EDITORIAL

EXECUTIVE EDITOR
Mike Senese
mike@makermedia.com

SENIOR EDITORS
Keith Hammond
khammond@makermedia.com
Caleb Kraft
caleb@makermedia.com

EDITOR
Laurie Barton

PRODUCTION MANAGER
Craig Couden

CONTRIBUTING EDITORS
William Gurstelle
Charles Platt

CONTRIBUTING WRITERS
Eduardo Alarcón, Debra Ansell, Joe Barnard, Guido Bonelli, Matthew Borgatti, Sam Brown, Spencer Chlebina, Sandra Ibrahim, Bob Knetzger, Elan Lee, Josh Lewis, Kari Love, Poppy Mosbacher, John Pedersen, Eric Steuer, Ted Tagami, Bram Torrekens, Sophy Wong, Melodie Yashar, Mendy Yu

DESIGN & PHOTOGRAPHY

ART DIRECTOR
Juliann Brown

MAKEZINE.COM

WEB/PRODUCT DEVELOPMENT
Rio Roth-Barreiro
Alicia Williams

ONLINE CONTRIBUTORS
John Allwine, Steve Altemeier, Jennifer Blakeslee, Kathy Ceceri, Chiara Cechini, David Cole, Ian Cole, Royce Florian, Jeff Gabel, Gretchen Giles, Goli Mohammadi, Winston Moy, Ryan Priore, Alvaro Jara Rodelgo, Adora Svitak, AnnMarie Thomas, Danielle Zimmerman

PARTNERSHIPS & ADVERTISING
makermedia.com/contact-sales or partnerships@makezine.com

SENIOR DIRECTOR OF PARTNERSHIPS & PROGRAMS
Katie D. Kunde

DIRECTOR OF PARTNERSHIPS
Shaun Beall

STRATEGIC PARTNERSHIPS
Cecily Benzon
Brigitte Mullin

DIRECTOR OF MEDIA OPERATIONS
Mara Lincoln

DIGITAL PRODUCT STRATEGY

DIGITAL COMMUNITY PRODUCT MANAGER
Matthew A. Dalton

MAKER FAIRE

MANAGING DIRECTOR
Sabrina Merlo

COMMERCE

OPERATIONS MANAGER
Rob Bullington

PUBLISHED BY
MAKER MEDIA, INC.
Dale Dougherty

Copyright © 2019 Maker Media, Inc. All rights reserved. Reproduction without permission is prohibited. Printed in the USA by Schumann Printers, Inc.

Comments may be sent to:
editor@makezine.com

Visit us online:
make.co

Follow us:
🐦 @make @makerfaire @makershed
f makemagazine
makemagazine
makemagazine
twitch.tv/make
makemagazine

Manage your account online, including change of address:
makezine.com/account
866-289-8847 toll-free in U.S. and Canada
818-487-2037,
5 a.m.–5 p.m., PST
cs@readerservices.makezine.com

CONTRIBUTORS

If you were boarding a ship to Mars, what's one thing you'd make sure to bring and why?

Debra Ansell
Los Angeles, California
(Edge-Lit LED Rainbow)

I'd bring a copy of Ray Bradbury's *The Martian Chronicles* to remind myself of the mystery, awe, peril, and responsibility inherent in exploring new worlds.

Josh Lewis
Aylett, Virginia
(DIY Cyanotype Camera)

If I were boarding a trip to Mars I would assume there were already cameras for the mission, so I would just pack a warm pair of socks.

Poppy Mosbacher
Brighton, England
(DIY Kaleidoscope Fabric)

I'd take a 3D printer to make things we didn't expect to need on Mars.

Ted Tagami
Berkeley, California
(Experimental Thinking)

I'd bring along a greenhouse with water and nutrients. Not only could we grow tasty food, but Mars could use a little green!

Joe Barnard
Nashville, Tennessee
(Fly Like SpaceX)

I live in Nashville, so I'd have to bring a solid supply of hot chicken!

Issue No. 69, June/July 2019. *Make:* (ISSN 1556-2336) is published bimonthly by Maker Media, Inc. in the months of January, March, May, July, September, and November. Maker Media is located at 1700 Montgomery Street, Suite 240, San Francisco, CA 94111. SUBSCRIPTIONS: Send all subscription requests to *Make:*, P.O. Box 17046, North Hollywood, CA 91615-9588 or subscribe online at makezine.com/offer or via phone at (866) 289-8847 (U.S. and Canada); all other countries call (818) 487-2037. Subscriptions are available for $34.99 for 1 year (6 issues) in the United States; in Canada: $39.99 USD; all other countries: $50.09 USD. Periodicals Postage Paid at San Francisco, CA, and at additional mailing offices. POSTMASTER: Send address changes to *Make:*, P.O. Box 17046, North Hollywood, CA 91615-9588. Canada Post Publications Mail Agreement Number 41129568. CANADA POSTMASTER: Send address changes to: Maker Media, PO Box 456, Niagara Falls, ON L2E 6V2

PRINTED WITH SOY INK

Kudos, Kids, and Corrections

MAKE: IS FOR THE KIDLETS

Sarah Hodsdon
@sarahndipitous

Following ∨

@Make inspires the very best out of #YoungMakers ... My family and Kidlets are better humans b/c of all the crazy amazing innovative people (and their creations) #Make has introduced us to over the years... HIGHLY recommend

Make: ✔ @make
@sarahndipitous teaches kids how to robot!
Show this thread

12:44 PM - 11 Mar 2019

SHORT, BUT SWEET PRAISE FOR MAKE: VOL. 68

Good issue, and best board guide yet — hit all the important and current things. —*Phillip Torrone, New York City, New York*

MAKE: AMENDS

I'm an electronics hobbyist, and I look forward to the annual "Boards" issue of *Make:* magazine. For the past couple of years, Digi-Key has sponsored a pullout "Guide to Boards" supplement in that issue (it's the April/May issue this year). And for the past couple of years, that guide has contained an error. My guess is that it just gets copied and pasted from year to year, so the mistake persists.

It's in the specs of the Adafruit Feather Huzzah microcontroller board — page two of the supplement, fourth line in the table.

Under the "Radio" column, it's listed as having both Wi-Fi and Bluetooth. But the Adafruit Huzzah uses the ESP8266, which only has Wi-Fi — no Bluetooth.

Adafruit does make a similar board, the Feather Huzzah32, which uses the ESP32 module. *That* one does have Bluetooth. The regular Huzzah does not.
—*Eric Yeater via email*

A NEW WAY TO READ MAKE:

Some eagle-eyed Twitter users saw a familiar title in the Apple News Plus announcement. (OK, Executive Editor Mike Senese may have helped it along.)

Mike Senese ✔
@msenese

Following ∨

Spotted this. We're part of the pitch.

10:30 AM - 25 Mar 2019

Chris Anderson ✔
@chr1sa

Following ∨

Replying to @WiredUK

There's @make! 🖤

10:21 AM - 25 Mar 2019

3 Likes

♡ 3

John Abella
@johnabella

Following ∨

Nice seeing @make in today's Apple News+ intro! Awesome. #applenews

10:15 AM - 25 Mar 2019 from Pennsylvania, USA

2 Retweets 7 Likes

♡ 1   ⟲ 2   ♡ 7

Christopher Urbanski
@ChrisUrbanski

Follow ∨

$9.99 for just @Make Magazine on #AppleNewsPlus is worth it by itself.

1:51 PM - 25 Mar 2019

2 Retweets 3 Likes

♡   ⟲ 2   ♡ 3

Apple

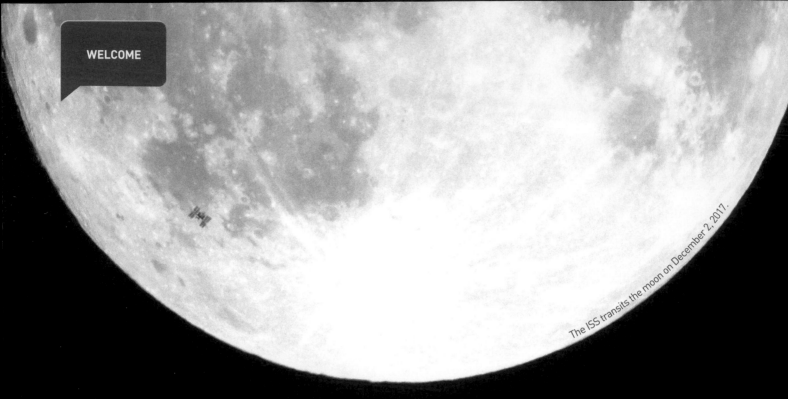

The ISS transits the moon on December 2, 2017.

15 Years of Make:

BY DALE DOUGHERTY, founder and CEO of *Make:* magazine and Maker Faire

One piece of advice I was given when I started out to develop *Make:* magazine was to be sure that we had plenty of good project ideas — and that we wouldn't exhaust those ideas after a few issues. As it turned out, finding good projects hasn't been a problem because we tapped into a rich supply of makers who were creating them. Once we started, the ongoing discovery of makers was made easier as more of them began to share with us what they were doing.

As I reflect back on the 15 years of *Make:* (and 14 years of Maker Faire), I am most grateful that we find makers of all stripes everywhere all over the world. The launch of this magazine in 2005 laid the groundwork for the global maker movement to emerge. We recognized makers for who they are: enthusiasts. They are bright, creative, and clever. In this magazine, we shared the instruction set behind maker projects so that others could learn from them and apply that learning to creating projects of their own.

We have also learned to recognize the special mindset of makers, and how hands-on learning changes us and can change the world. We seek to spread the practice of making throughout society, especially through all levels of formal and informal education. (A special shout out to teachers and librarians who have become champions of the maker movement.) The good news is that makers are getting younger and younger every year. The world needs more of them and that has become our mission.

As I look around me in 2019, I know that "maker" has become a widely used term. However, I also see ever more reasons for the maker community to widen its impact. In this time when our culture is polarized, when many people are angry and depressed, when companies and workers wonder what the future holds for them, this kind of cultural expression based on play, creativity, and resilience is an antidote for what ails us. It remains our best hope for imagining and creating the kind of world we want our children to live in.

Fittingly, this issue is about space, which coincides with the 50th anniversary of humans landing on the moon. It is our 69th regular issue (plus 8 special issues over this period.) Thanks to all the contributors to *Make:* who have shared their how-to projects. Thanks to our dedicated editorial and creative teams for their work producing each issue. A special thanks to our loyal readers whom we have always called makers from the very beginning. All of you have made this moon shot possible.

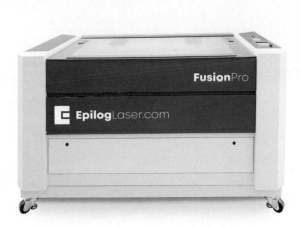

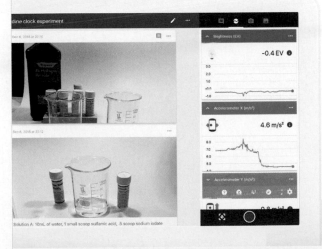

MADE ON EARTH

Backyard builds from around the globe

Know a project that would be perfect for Made on Earth? Let us know: *editor@makezine.com*

IN GOOD HANDS

TALAVN.COM.VN

"I want people to feel like they're walking in the clouds," says **Vũ Việt Anh**, the architect behind *Cầu Vàng*, a breathtaking footbridge in the Bà Nà Hills resort of central Vietnam. Also known as *The Golden Bridge*, the structure sits nearly 3,300 feet above sea level, bolstered by a pair of massive supports that are fashioned to look like aged hands made of stone. The appendages, which were actually constructed with fiberglass, mesh wire, and a steel core, appear to be reaching out of the ground to lift pedestrians high up into the sky. Meanwhile, the walkway resembles a thin strand of gold. It's bordered by side rails that curve inward to give the impression that parts of the bridge are set at a steep sideways tilt.

The bridge, which is nearly 500 feet long and includes eight spans, opened last summer as part of a $2 billion campaign to attract tourists to the area just outside the city of Da Nang. Anh and his firm, TA Landscape Architecture, were initially commissioned to build a simple connection between two cable car stations, but ultimately opted for a much more ambitious design that would grant visitors a birds-eye view of the mountains and sprawling gardens that surround the resort. "Our inspiration came from nature," Anh says, noting that he imagined what a bridge in heaven might look like when he and his team created *Cầu Vàng*'s mock-ups.

Anh says that *Cầu Vàng* took almost a year to construct and requires frequent maintenance — due in large part to the fact that it has become a social media sensation, flooded daily with visitors eager to snag selfies on the world's coolest new bridge. Indeed, a search on Instagram for the hashtags #goldenbridge and #cauvang turns up tens of thousands of stunning photos. —*Eric Steuer*

FUNCTIONAL FREEDOM

PABLOREINOSO.COM

For the past few decades, Argentine-French sculptor **Pablo Reinoso** has been crafting artful furniture that imbues traditional (and even mundane) functional design with surrealistic flourishes inspired by the natural world. Perhaps the best example of this approach is the artist's *Spaghetti Bench* series, launched in 2006, which reimagines the humble park bench — a design so common and so enduring that you have likely never stopped to think of its origins — as a starting point for limitless creative possibilities. The wooden slats in Reinoso's benches twist and turn in wild ways, reminding us that they began as living things that grew in beautiful, unpredictable directions.

While Reinoso exhibits at some of the most high-end art galleries and exhibitions around the world, he is at heart a craftsman. Introduced to carpentry by his grandfather, Reinoso began designing chairs when he was just 6 years old, and has been creating experimental furniture by hand since his teens. Continuing in that tradition, each of the pieces of wood that make up his intricate *Spaghetti Bench* designs are sculpted, not shaped or manufactured. As a result, the work carries with it an unmistakable warmth — a sense that it is just as dynamic as the plant life that provided its source material. —*Eric Steuer*

ENGINEERED ART

BHOITE.COM

LEDs, ICs, and sliders stand suspended on various sides of an open cube. Thin metal rods hold sensors apart from rows of LEDs like a tiny racetrack starting light. Seven-segment displays rise up and tilt away from controller boards. These peculiar structures almost appear to be exploded views of circuits, with metal leads extending cleanly between components like annotations on a diagram. And while they're intriguing to look at, they're actually fully functional too. Created by **Mohit Bhoite**, these pieces are the culmination of years of obsession with "dead bug"-style circuitry — circuits soldered together without a PCB to hold everything in place.

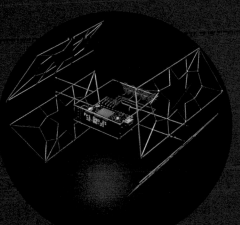

His first creations came around 2004; they were well received, but he set aside the hobby for a while. When Bhoite revisited the idea in 2018, he found that people all over the world were inspired by his work. "The reactions have been extremely pleasant and encouraging," he says. "I started sharing pictures of the projects on Instagram, which has led to a big following. I get a lot of messages from young makers and students who really want to get into electronics so that they could build this."

Bhoite, a circuit designer at Particle, draws inspiration for his pieces from Japanese carpentry, truss bridges, Islamic patterns, and industrial plumbing. He admits to enjoying the challenge of thinking about his electronics in three dimensions. "As engineers we have been trained to solve problems efficiently, but when you mix in art, you have to unlearn some of that a little bit to achieve the aesthetics that you are looking for," he says. "It's a fun and challenging space to work in when your engineer brain and art brain have to collaborate." —Caleb Kraft

Mohite Bhoite

ROAD WARRIORS

GRAYWRX.COM

Built from scrap steel, discarded parts, antique machinery, and industrial equipment, the robots of **Justin Gray**, a self-described "half artist, half engineer," are unique and full of life. While they may just be machines piloted remotely by people, their creative construction gives each the feel of a personality all its own. Some appear harmless and even cute while others exhibit a lust for destruction with spewing flames and spinning chainsaws. Many use heavy metal treads for locomotion; power can come from batteries, hydraulics, and even turbine engines.

Gray has been delighting the crowds with his creations at Maker Faire since its beginning in 2006, while his work dates back nearly a decade before that. But it was an early international incident that helped form the desire to build remotely controllable robots that's become a big part of his life. "I was 10 years old when the Chernobyl accident happened," Gray says. "Seeing the robots used instead of humans was a key moment."

Based in Oakland, Gray is also co-founder of Cooper Gray Robotics (robotloader.com), building remote operated electric drive vehicles used in construction and other material-handling-dependent industries. "Picture a skid steer loader, but dead silent and no operator, built sealed tight for harsh environments," he says.

With a personal portfolio of eight creations, Gray has more than enough to keep his hands full, but still finds time to do side projects, including building electric vehicles, producing special effects and flame work, and constructing a 12-ton amphibious-track-drive fire-fighting robot for Jamie Hyneman. "It is driven using a remote interface inside a VR environment using the Xbox engine and Oculus Rift." —*Caleb Kraft and Mike Senese*

Justin Gray

SIN CITY CLEANS UP

WRITTEN BY ERIC STEUER

ERIC STEUER is the creative director of Creative Commons and the producer of *Plays Well with Others*, a podcast about the art, science, and mechanics of collaboration.

SEVEN YEARS IN, THE DOWNTOWN LAS VEGAS PROJECT – AND A BEVY OF MAKERS – **HAVE TRANSFORMED A FORGOTTEN PART OF THE CITY**

WITHIN SECONDS OF ENTERING THE COURTYARD AT FERGUSONS IN DOWNTOWN LAS VEGAS, I meet Marley. I jump back a bit when I realize that the animal I'd noticed only peripherally and assumed was some kind of big designer dog is actually a black alpaca. He's friendly as can be, with a sweet smile that I choose to interpret as an expression of fondness, even though I know deep down it's just how alpacas' faces always look.

Danielle Kelly is Fergusons' property manager. She takes me to where I'll be staying for the night: an Airstream trailer, a tidy and top-of-the-line model with several small rooms inside, two TVs, and a nicer sound system than I have at home. It's one of a few dozen Airstreams and tiny houses that sit on the property, part of an effort to bring creative entrepreneurs together for an experiment in living communally and with minimal possessions.

"And here's your neighbor," Kelly says, signaling to look into a window of a unit near mine. I squint to make sure I'm really seeing what I think I'm seeing. "Is that ... a sloth?,"

I ask. Kelly nods. I already know the answer to my next question, but I ask it anyway: "And he lives in his own little house?"

Kelly tells me that Airstream Village, as it's known colloquially, has been up and running for a few years, but was just recently relocated to a new home — here, the former site of Fergusons Motel, one of the original lodges that lined Fremont Street back in Downtown Las Vegas' old days. There are big plans in store for Fergusons. In addition to managing the residential community and their humble abodes, the team here is sprucing up the old inn's rooms with a spiffy vintage motif in anticipation of reopening them to travelers. Cool new retail spots are going in soon. A skate park is being built. There will be lots of art — *big art*. Out front, there's the 42-foot-tall sculpture, *Big Rig Jig*, made from a pair of 18-wheelers.

Downtown used to be the fashionable part of the city. If you're thinking of Rat Pack-era Vegas, you're thinking of downtown. As the bustle steadily moved away to the Strip in the '80s and '90s, Vegas' original hot spot was neglected and over time became blighted. The old Fergusons sat more or less abandoned for several years, until it was purchased by Downtown Project, the urban

revitalization endeavor launched in 2012 by Zappos CEO Tony Hsieh after he sold the online shoe and clothing retailer to Amazon for $1.2 billion.

• • •

If you've heard much of anything about Downtown Las Vegas over the past few years, you're almost certainly aware of Downtown Project. Rebranded as DTP Companies late last year, the $350 million initiative is focused on creating a new type of community-centric neighborhood, shaped in large part by a commitment to innovators and makers. Over nearly 45 acres in downtown's Fremont East district, DTP operates a few dozen of its own businesses and creative ventures in the neighborhood, and either invests in or partners with several dozen others.

At the project's heart has always been a desire to see what happens when you let makers run a city in their vision — even (or perhaps especially?) when that vision includes ideas that might seem a little wacky to outsiders. DTP has had its share of setbacks (several of its early investments failed) and detractors (folks who argue

NeonPR, Eric Steuer

"AND WE TOLD OURSELVES THAT IF WE EVER HAD THE CHANCE, WE'D DITCH THE WORKADAY WORLD AND FIND A WAY TO BRING INTERESTING PEOPLE TOGETHER SOMEWHERE WEIRD TO FOCUS ON BEING CREATIVE."

that despite all the money and work that's gone into re-envisioning the district, it's still more of a hipster wonderland than a "real city"). But there's a renewed energy here as of late, a second wave of activity and development designed to bring creative folks to town and keep them here for the long haul. In addition to undertakings like Fergusons, a host of housing is being developed to handle new arrivals eager to be a part of the action.

Complementary efforts like the City of Las Vegas' own Project Enchilada, an expansive plan concentrated on restoring vintage architecture and bringing new landscaping to the area, are also taking hold.

While Hsieh is still heavily involved in DTP, he stepped down from his formal leadership position back in 2014. Michael Downs, DTP's executive vice president, is now responsible for driving the effort's big-picture planning and its daily operations. "The primary objective is to bring more bodies to live, work, and play in DTLV," Downs says, citing the many projects under the DTP umbrella, which all share an executive board and a

central marketing team.

One thing is certain: Downtown Las Vegas has undergone a remarkable amount of change since the DTP team first took it on. I've visited nearly 20 times over the past decade, and it would be hard to overstate how different it feels now versus 10 years ago, before the revitalization kicked off. I'm always amazed by how quickly and how often things pop up — every time I'm in town, I encounter new and inspiring art, technology, and business ventures.

And I always meet interesting people. Today, one of those people is Jen Taler, the curator behind Market in the Alley, a monthly showcase for local makers ranging from woodworkers to chefs. "We created this because we saw a need for people to have a way to share all the cool stuff they're doing with the rest of their community," she says. "It's also a space for people who have ideas for companies that they might eventually want to level up. I think of it as an incubator in a way." She introduces me to some of the market's exhibitors — a resin artist who makes custom kitchenware, a couple who sell their own line of vegan hot dogs (Las Vegas' first), and a group called Future Makers Las Vegas, which leads

crafting, gardening, and entrepreneurial workshops for kids. Nearly 1,500 attendees make their way into the market over the course of the day.

• • •

Later that afternoon, I take a walking tour of the neighborhood. My first stop is Container Park, an outdoor shopping center built out of repurposed shipping containers and Xtreme Cubes. Several dozen small businesses reside there, all of them local. They surround a playground with a three-story treehouse and a large, electronic play structure programmed with interactive games for kids. Out front is The Dome, a 360° immersive theater that features a 14-million-pixel 4K display. And then there's the Praying Mantis, a 55-foot-tall steel sculpture with a dump truck at its base. Built by artists Kristen Ulmer and Kirk Jellum as an exhibit for Burning Man, it's designed to "dance" and breathe fire in rhythmic bursts when music plays at night. When I return later in the evening, it's gettin' down to James Brown's "I Feel Good."

I stroll through the Arts District and take photos of huge outdoor murals and

installations by world-renowned artists like Pixel Pancho, Ana Maria Ortiz, Shepard Fairey, Mark Drew, and D*Face. Many of the works were commissioned by the DTP team that runs Life Is Beautiful, the music and culture festival that's been held annually in the neighborhood for the past five years. My favorite piece from the collection is André Saraiva's *The Empty Club* — a mock nightspot composed of a small pink and blue graffitied building (formerly the office of a deserted gas station) with a lonely mirror ball hanging from the ceiling inside.

I stop by the Gold Spike, a casino that DTP bought and converted for daytime use as a drop-in workspace. I hop online and a woman asks me if I'm there for the hacker meet-up. When I walk outside, I watch a guy in full Elvis costume argue with a guy dressed like a robot about where they're going to set up on the street to perform.

On my way back up to Fergusons, I pass the Hydrant Club, a dog care and training center with a large open area in front for canines and their humans to commune. It's home to a 15-foot-tall functioning fireplug that lights up from the inside. A group of women stand near it, taking selfies. "It's so crazy to see all this stuff here," one of them says. "When I was a kid, this was a part of town you just *did not go*."

• • •

I've never been able to quite put my finger on what it is about Downtown Las Vegas that's so attractive to me. While it grows and changes and gets more fun each time I'm there, it's also far from perfect. There are still significant stretches that remain desolate and dilapidated, and it feels like it's still missing critical infrastructure that would be necessary for a permanent community to really congeal. But after this last trip, after meeting some of the people who are instrumental in designing the next phase of the neighborhood's plan, I've figured out what my fascination stems from. It's that the creative energy that DTP launched with, and that remains in full force seven years later, reminds me of many excited conversations I had with friends way back when we were young and idealistic. We complained to each other about how all people with money care about is turning their money into even more money. We told ourselves that if we ever had the chance, we'd ditch the workaday world and find a way to bring interesting people together somewhere weird to focus on being creative. We knew this was naive and probably impossible. But it was fun to talk about.

When I'm downtown, I feel some of that excitement again. And I wonder: If you did have the resources and the opportunity, why wouldn't you find some neighborhood with myriad creative possibilities, a place that just needs some love and attention and people willing to make some things happen — and make some things happen? Why wouldn't you invest in artists who make big, strange, audacious stuff for no other reason than because it's possible to do, and then install their work on street corners around town? Why wouldn't you create a place for people who want to do things like start small craft businesses or serve up specialty food or teach kids to build things? Why wouldn't you turn the grounds of an expired motel into a high-tech trailer park where residents live alongside animals that most people only see at the zoo? Seriously, why not? ●

Fur Sale

Written by Elan Lee

How an **Exploding Kittens** vending machine sparked convention joy, one random item at a time

ELAN LEE is a professional technologist and storyteller. He founded several companies including 42 Entertainment and Fourth Wall Studios. He was the Chief Design Officer at Xbox Entertainment Studios, and co-created Exploding Kittens, the most funded game on Kickstarter, and the most backed crowdfunded project in history. His latest project is Throw Throw Burrito, a dodgeball card game.

CRUNCH
MUNCH
MUNCH

IN 2015, WE LAUNCHED A KICKSTARTER CAMPAIGN FOR A GAME CALLED EXPLODING KITTENS.

The game did really well, and the campaign broke a bunch of records. But this isn't a story about a four-year-old card game, it's a story about joy, and magic ... and a really massive machine we built that nearly got us kicked out of every convention we brought it to.

To run a successful game company, you have to attend conventions; these are some of the best opportunities for brands to get face time with their fans, and not only stay relevant in the overall ecosystem, but hopefully generate buzz. In addition to San Diego Comic-Con, there's Gen Con, Dragon Con, WonderCon, KittyCon, Ohdearlordnotanothercon ... Actually, at least one of those is fake, but you get the idea. And by the time this goes to print, it will probably be a real event. People love cons.

other peoples' plastic toys, cookie-cutter games, and utterly forgettable booths. We accomplished nothing. By the time we started prepping for year two, we knew we had to come up with something better.

BUILDING A BETTER MOUSETRAP ... ER, CONVENTION BOOTH

We started deconstructing the notion of a "booth."

1. What does a transaction look like?
   Give money, get product.
2. How long is each interaction?
   20 seconds.
3. What are the physical attributes?
   Audience-facing attractors, sample products displayed, prices listed, hidden inventory.

Basically, every booth is a vending machine but designed by someone who doesn't realize they're building a vending machine. To succeed at conventions, we realized, we had to build the world's

> To succeed at conventions, we realized, we had to build the world's best vending machine. So we did.

These events are huge. They are noisy, crowded with thousands of fans, and packed with hundreds of game makers clamoring for 20 seconds of your attention.

The first year we went to Comic-Con, we did what we were supposed to. Our 2016 booth looked just like everyone else's, and we got buried beneath a torrent of

best vending machine. So we did.

We started with a giant cardboard box — and then covered it in fur. We made sure our machine had all the features that a perfect vending machine needed: buttons to pick a product, a place to insert money, a chute for product delivery, and a giant screen to interact with customers.

Exploding Kittens

Most importantly, this machine had a very special and magical secret weapon — a rainbow-colored button that simply read "Random Item $1."

Whenever that button was pushed, a random item was delivered quickly through a slot in the machine. Items included whole watermelons, toilet plungers, custom drawn artwork, brooms, a bag of rocks, sombreros, or asparagus.

The random items were designed to entertain, amaze, and delight every audience member brave enough to push the button.

So what was the secret to this impossible technological marvel? How did we do it?

We filled it full of people.

THE POWER OF RANDOM STUFF

The Exploding Kittens vending machine was actually a vending machine costume built for our team. In addition to all the products and merchandise, the machine contained thousands of different random items, and a team of six to 12 people working nonstop to turn these items into a seemingly infinite number of possibilities. Audiences could watch this machine for hours and never see the same item twice.

After the first day of running the machine we had underestimated demand for random items so drastically that we had to make an emergency run to the local dollar store and buy every item we could fit in our cars. We had expected to deliver 250 random items during the convention. We ended up delivering 1,400. That's 40 items per hour for 35 hours over the course of the weekend.

Fans lined up for hours spilling over into

walkways, hallways, and other people's booths. The lines got so long that we attracted the attention of the fire marshal, which it turns out is a great way to almost get kicked out of conventions. We had to hire line monitors, and whenever the wait exceeded three hours, just had to turn people away.

It would be difficult to overstate the results our machine produced. Every moment was a spectacle. The machine took selfies with Storm Troopers, it dressed kids up in fairy wings and unicorn horns, and even performed magic tricks.

We had to learn a lot in a very short amount of time. We learned that balloon animals were a huge crowd favorite, and so we had to frantically watch YouTube videos backstage to learn how to make them. We learned that in many states you have to be a registered grocer to have produce delivered to a convention hall, so Exploding Kittens is now a registered grocer in 14 states. We learned that we could interact with the audience on a personal level, like when we spotted someone dressed as Daenerys Targaryen (the Mother of Dragons from *Game of Thrones*) in line, we bedazzled an entire watermelon so that when she got to the front of the line, we could instantly deliver a dragon egg and blow up her brain.

We worked so hard to play with the crowd, and they played right back.

They wrote love notes to the machine, and one person actually proposed marriage (for which she received a massive bouquet of 50 red roses). People shared their stories all over the convention floor, in the hallways, and on social media. The machine was one

of the biggest hits at every convention we brought it to.

The final result was pure magic: a live show that crafted impossible and unique moments for each individual in real time every single minute, nonstop, for 8 hours a day.

People laughed, people cried, people fell in love, and people made friends with this machine.

Our goal was to stand out at conventions, and we ended up building a wondrous joy machine running on human software and vending happiness.

This was not just a machine that solved our convention booth problem — it was the best machine we've ever built. ⊘

> It would be difficult to overstate the results our machine produced. Every moment was a spectacle.

Exploding Kittens

Written by Sandra Ibrahim

AROUND THE WORLD

NO MATTER WHERE YOU ARE, THERE'S A **MAKER FAIRE** HAPPENING NEAR YOU

EVERY YEAR THE WORLD IS FILLED WITH EXCITING MAKER FAIRES happening around the globe, and 2019 is no exception. Maker Faire is unique because of the makers, the locations, and venues. But it's each event's producers and team leaders that help create the magic behind the scenes. The producers have the crucial role of organizing all aspects of their shows for their attendees, giving a platform to the makers to showcase their projects and inventions, and sharing them with both their local communities and the world.

There are over 150 Maker Faires happening this year, each unique and incredible. Here are five shows you should put on your calendar, ranging from second-year events in Prague, Glasgow, and Guadalajara, to a decade-old institution in Detroit. Be sure to visit makerfaire.com/map to find even more.

1. MAKE FAIRE GLASGOW

Location: Riverside Museum & Tall Ship, Glasgow, Scotland
Date: Autumn 2019
Annual: 2nd
Producers: Aziz Rasool, Jim Watt, Kim Scott, and Martin Goodfellow
Aziz Rasool and his group Artronix are producers of Maker Faire Glasgow, appearing in its second year this fall. They launched the event to showcase diverse makers from the area, as well as the region's heritage of innovation, arts, design, and science. Rasool points out Hacky Racers — a group that soups up kid's

electric cars — and the experimental music and technology group Sensatronic-Lab as a couple of favorites appearing this year. The event will be held at the Riverside Museum on the Clyde River; nearby are the Glasgow Science Centre and GalGael, an incredible community workshop where people come to practice historic woodworking and boat building.

2. MAKER FAIRE DETROIT

Location: The Henry Ford Museum, Detroit, Michigan
Date: July 27–28, 2019
Annual: 10th
Producers: Cynthia Jones, Jim Johnson, and Kristi Best
The Henry Ford Museum has a couple things to be proud of this year — they are having their 90th anniversary and are producing the 10th annual Maker Faire Detroit, hosted onsite. As producer Cynthia Jones points out, "Those are two cool milestones to mark together." For the past 19 years, Jones has co-created content and programs through her work as a "museum geek;" she's also a 22-year Burning Man attendee and is the co-founder of the Great Lakes Regional Burning Man event, Lakes of Fire. Jones used these experiences to help launch a great installment of Maker Faire inside a venerable location. "We saw the inspirational power of having a Maker Faire inside of a museum," she says, "juxtaposing cutting edge inventions of today with marvels of yesterday."

3. MAKER FAIRE TOKYO

Location: Tokyo Big Sight, Tokyo, Japan
Date: August 3–4, 2019
Annual: 8th
Producers: Hideo Tamura, Fumi Yamakawa, and Yuko Asahara
Maker Faire Tokyo is curated by Hideo Tamura, an editor at O'Reilly Japan. Initially working on the Japanese edition of *Make:* magazine, Tamura learned about the various Faires around the world. Their first event ("a huge success") had 30 makers and 600 visitors. It's been upward from there; last year's Tokyo Faire — their 7th annual — had grown to a giant 600 makers and 24,000 visitors, with many hands-on activities and workshops.

Tamura says their Faire is unique because of "the Japanese sense of humor, the details of the projects, and the making of things as a hobby from the long historical background of the Japanese makers."

4. MAKER FAIRE PRAGUE

Location: Výstaviště Praha, Prague, Czech Republic
Date: June 22–23, 2019
Annual: 2nd
Producers: Jiri Zemánek, Denisa Kera, Ivan Sobička, Leona Daňková, Jasna Sýkorová, Daniel Jirotka, Ondřej Kašpárek, Vojta Kolařík, and Martin Pikous
Prague's Jiri Zemánek came to Maker Faire Bay Area a few years ago and was immediately enamored with the idea of producing an event in his own city. He and his team finally made it happen in 2018,

SANDRA IBRAHIM helps coordinate the Global Maker Faire program at Maker Media.

hosting their show in a gorgeous glass-and-steel palace that dates back to 1891. Last year's event brought together exhibits including a giant air cannon, homemade Nixie tubes, and a maker creating working steam engines from glass. It was so popular that the producers are expanding the maker movement to two other Czech cities this year, Pilsen and Brno. "DIY culture has a long tradition in the Czech Republic," Zemánek says, "although in the past it was motivated mostly by lack of things behind the Iron Curtain."

5. MAKER FAIRE JALISCO

Location: Centro de la Amistad Internacional / Expo Guadalajara, Mexico
Date: November 8–9, 2019
Annual: 2nd
Producers: Darius Lau Castro, José Ruiz, Oscar Pereda, Hans Ramirez, Yizia Jimenez, Joel Gastelum, and Eli Salmeron

Maker Faire Jalisco will be the first "Featured" Faire in Mexico, organized by the volunteer group MakersGDL. Producer Darius Lau Castro has been coordinating maker events with this team for the past five years, from meetups to hackathons to conferences. He says, "Mexico is a joyful country by nature and we have our own traditions when it comes to faires … food, games, colors, and a DIY mechanical attractions tradition. We want to take those traditions and re-imagine the Mexican Faire with the maker spirit." ◎

EVERY FAIRE, EVERYWHERE

• Akron, OH • Alameda, CA • Albuquerque, NM • Asheville, NC • Ashland, OR • Atlanta, GA • Aurich, Germany • Austin, TX • Bangkok, Thailand • Barcelona, Spain • Baton Rouge, LA • Beirut, Lebanon • Berlin, Germany • Biddeford, ME • Bitola, Republic of North Macedonia • Bloomsburg, PA • Bonifacio Global City, Philippines • Bowling Green, KY • Brno, Czech Republic • Brussels, Belgium • Burlington, NC • Cairo, Egypt • Canton, OH • Casablanca, Morocco • Charlotte, NC • Chemnitz, Germany • Chicago, IL • Cincinnati, OH • Coeur d'Alene, ID • Colorado Springs, CO • Columbia, SC • Corning, NY • Corona, Queens, NY • Cranberry Township, PA • Darmstadt, Germany • Dearborn, MI • Des Moines, IA • Dortmund, Germany • Edmonton, Alberta, Canada • Eindhoven, Netherlands • Elkhorn, WI • Evorá, Portugal • Fairfax, VA • Gainesville, FL • Ghent, Belgium • Glasgow, Scotland • Göttingen, Germany • Grand Rapids, MI • Guadalajara, Jalisco, Mexico • Hamburg, Germany • Hannover, Germany • Hattiesburg, MS • Herford, Germany • Hong Kong, China • Honolulu, HI • Hyderabad, India • Idaho Falls, ID • Jacksonsville, FL • Kalispell, MT • Kansas City, MO • Kent, OH • Kingsport, TN • Kyoto, Japan • Lafayette, IN • Lafayette, LA • Lakeville, CT • Lehi, UT • Lille, France • Ljubljana, Slovenia • Loma Linda, CA • Los Angeles, CA • Louisville, KY • Lynchburg, VA • Madison, WI • Madrid, Spain • Martinsville, VA • Mashpee, MA • Meridian, MS • Miami, FL • Minden, Germany • Milwaukee, WI • Mishref, Kuwait • North Little Rock, AR • Northampton, MA • Nürnberg, Germany • Oakland, CA • Orlando, FL • Orrville, OH • Palm Bay, FL • Paris, France • Pennington, NJ • Pilsen, Czech Republic • Port Jefferson, NY • Portland, Oregon • Prague, Czech Republic • Providence, RI • Riyadh, Saudi Arabia • Roanoke, VA • Rochester, NY • Rocklin, CA • Rome, Italy • Rosport, Luxembourg • Saint Joseph, MO • San Antonio, TX • San Diego, CA • San Jose, CA • San Mateo, CA • Santiago de Compostela, Spain • Saskatoon, Saskatchewan, Canada • Schenectady, NY • Seoul, Korea • Sheboygan, WI • Shelburne Farms, VT • Shenzhen, China • Shreveport, LA • Simsbury, CT • Skopje, Republic of North Macedonia • South Bend, IN • Springfield, MO • St. Joseph, MI • Tacoma, WA • Taipei City, Zhongzheng District, Taiwan • Tampa, FL • Tokyo, Japan • Torino, Italy • Trieste, Italy • Trondheim, Norway • Tulsa, OK • Tyler, TX • Utica, NY • Vienna, Austria • Vista, CA • Wenatchee, Washington • West Tisbury, MA • Westport, CT • Wichita, KS • Windsor, Ontario, Canada • Wolfsburg, Germany • Zagreb, Croatia

MIKE SENESE is the executive
editor of *Make:* magazine.

SPACE GEEK

Written by Mike Senese

WITH SO MANY INCREDIBLE OPPORTUNITIES, THERE'S NO BETTER TIME TO BUILD SOMETHING FOR SPACE

"We have entered an era of accessible space. Space that is accessible to life-long learners, makers, and students. Getting tools and experiences into the hands of curious explorers today is not only fun and engaging, it will make for a better world tomorrow as we face significant challenges on our Earth habitat. A lot has changed since the beginning of space exploration. The profound changes occurring today become the opportunity for the dreamers, tinkers, thinkers, explorers, and makers of tomorrow.

"A big factor has been cost. For $3.4 billion (in 2019 dollars), the Surveyor missions from 1966-68 put five working spacecraft on the surface of the moon. This year a group of Israeli engineers launched the first privately funded mission to land on the moon for just over $100 million; it crashed upon descent but they remain undeterred. The trend is clear: While still difficult, space is substantially more affordable today. Welcome to the New Space Age.

"Significant changes in technology and accessibility mean that payloads that meet stringent NASA criteria can make their way to the International Space Station for under $50,000. This is a big deal. Eventually the moon, Mars, and beyond will be approachable as viable opportunities. The Billionaire Boys Club (Bezos, Branson, Musk), along with dozens of small launch vehicles, make the era of New Space a perfect set-up for 'Making Space.'

"And keep this in mind: Students in middle school today will be of astronaut recruitment age by the time we have humans on the Martian surface. Why wouldn't you want to make something for that?"

Ted Tagami sent me the above note as we engaged in a series of discussions around the exciting new developments in the DIY space arena to plan for this issue of Make:. As co-founder of Magnitude.io, an initiative that offers students access to space and near-space research opportunities, Ted is passionately plugged into this new space community, and loves bringing others along for the ride (more about Magnitude.io on page 44). Our conversations revealed new ventures happening on almost all sides of this community, from projects that bring the cosmos to the classroom, to startups and venerable players pursuing innovative and beneficial commercial endeavors. Takeaway: There are more ways now for anyone to get their projects orbital — or near to It — than ever before.

Here are some of the fascinating new ways you can connect to space, whether you're amateur or pro. And there are many more — we encourage you to explore and share your journeys too.

AMATEUR

ISS NATIONAL LAB

It's hard to imagine space research without the International Space Station. When the first ISS module settled into position 18

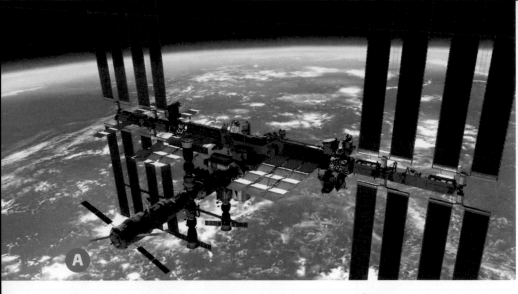

A

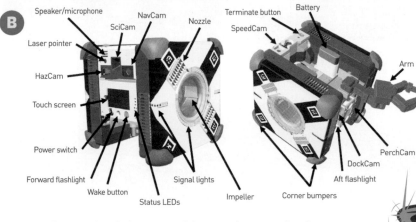

B

Speaker/microphone
SciCam
NavCam
Nozzle
Laser pointer
HazCam
Touch screen
Power switch
Forward flashlight
Wake button
Status LEDs
Impeller
Signal lights
Terminate button
SpeedCam
Battery
Arm
PerchCam
DockCam
Aft flashlight
Corner bumpers

years ago, it created a whole new model for accessing space. The ISS's commercial research aspects are countless, but perhaps more importantly, it's open for educational and public access.

In 2011, NASA awarded a multi-year contract to the Center for the Advancement of Science in Space (CASIS) to manage the ISS National Lab, allowing them to implement educational and commercial research in the lab. It's been a success. "We've launched hundreds of student experiments to the space station," says Dan Barstow, senior education manager for the ISS National Lab.

It's through the ISS National Lab that programs like Magnitude.io exist. Other activations include EarthKam (students set coordinates for aerial photos and receive the images after the ISS flies overhead; earthkam.org) and ARISS (classrooms speak with ISS astronauts via radio as they pass by; ariss.org). The lab approves half of the cargo that heads to the ISS through NASA launches, and gets half of the astronauts' time to operate the experiments

that comprise that cargo. And they're looking for more, especially from hobbyists. "In my opinion, the maker community is the biggest growth area that we should approach now," Barstow says. "The ISS is open for makers."

ASTROBEE AND ZERO ROBOTICS

Various robots have occupied the ISS through the years, including three floating, bowling ball-sized polyhedrons called SPHERES. MIT developed these semi-autonomous robots and in 2006 placed them on the space station "to study guidance and control, formation flight, in-space assembly, in-space construction," says Jose Benavides, NASA's project manager for the program. After years of use, they are now being replaced with a next-gen version called Astrobee (nasa.gov/astrobee), which will feature more tools and sensors, including display screens, a laser pointer, and a "perching arm" that extends out from one side and can grab onto items like handrails, to help the astronauts with

inventory and housekeeping — a vital need as we push our space outposts further into the solar system, including NASA's proposed moon orbiting station, Gateway.

Although the ISS robots are changing, one of their coolest programs will remain — a twice-yearly international student competition that requires middle- and high-schoolers to program the robots to navigate various challenges. Called Zero Robotics (a play on First Robotics, on which it is based; zerorobotics.mit.edu), the first few rounds of the competition are played using virtual robots in a simulated environment. The finalists then get to send their code to the ISS, where the astronauts load it onto the actual robots and have them enact the instructions in the real zero gravity environment to determine the winner.

The high-level Astrobee software is based on Android, and NASA has open-sourced the software stack and simulator (grab

A The ISS National Lab program has brought hundreds of student experiments to space.

B Astrobee, the ISS's new floating robot helper that students can program too.

C Rendering of Space Tango's proposed ST-42 unmanned orbital research station.

D ARISS and ISS-Above in use during a ham radio session between ISS astronauts and students at Kennedy Space Center.

E Bill Brown's tiny, solar-powered PicoHAB lets students track multi-day high-altitude balloon trips around the globe.

them at github.com/nasa/astrobee). "We want to provide as much help as we can to both students and academics so they can simply focus on their research," says Andres Mora, robotics and electrical lead on the Astrobee program.

SPACE TANGO

Putting research projects on the ISS requires some assistance, and that's where Space Tango (spacetango.com) has focused its energies since being established in 2014. The Kentucky-based group is a flight integration supplier to the ISS through the NASA Space Act Agreement. On board, it operates two automated cargo lockers, called TangoLabs, that accept and operate standardized research containers. This lets everyone from students to Fortune 500 companies conduct education and R&D projects in a simplified manner.

To expand those opportunities, Space Tango recently announced plans to launch its own unmanned, orbiting vessel, ST-42, in the mid-2020s, for microgravity research and even manufacturing. CEO

Twyman Clements says, "We have stem cell, organoid, chemical production, and other really interesting projects on our manifest."

ISS-ABOVE

With new maker-friendly tools, there are now low-cost ways to interface with space from your workshop or classroom. In 2013, software programmer Liam Kennedy designed and built a Raspberry Pi-based ISS notifier for his 3-year-old grandson Owen, to inspire him about space as the device blinked regularly. "The key thing about the space station, which I think often goes missed, is that it passes you by 5–8 times a day, somewhere in your sky," Kennedy says. Dubbed ISS-Above (issabove.com), the device started getting interest from others. Kennedy's local coffee shop, across the street from CalTech in Pasadena, asked if he could make them a unit too, then some friends suggested he show it at Maker Faire San Diego in 2014. This led to a successful Kickstarter later that year.

That same year NASA switched on a new capability on the ISS to broadcast live views

of Earth from orbit, which inspired Kennedy to update his device. "I very quickly realized this Raspberry Pi gizmo that I put together, I could reprogram it to also stream live video from the space station," he says. The second version of his device now does just that, plugging into a monitor to show live footage and other ISS information. There are now over 3,000 units in classrooms, offices, and elsewhere. (For more details see page 41.)

PICOHAB

Another great maker-friendly activity is sending a tracker to the edge of space via high-altitude balloon (HAB), a concept pioneered 30 years ago by NASA electrical engineer Bill Brown (wb8elk.com). Ever the tinkerer, Brown has continued advancing on the concept over the 500 launches he's done. "My latest projects have been to design PicoHAB payloads as lightweight as possible," he says about his trackers, which weigh in at a svelte 12 grams. "I had the PC board manufactured at about one-third the normal thickness, used a very lightweight brass tube from a modeling specialty store

CASIS, NASA/Ames Research Center/Dominic Hart, Space Tango, Liam Kennedy, Bill Brown

imagery they were able to see there were two villages that were not on any of the responders' maps," Safyan says. "They were able to deploy aid to places that would have likely been forgotten."

RELATIVITY SPACE

There are a number of new firms making commercial-grade rockets, most famously SpaceX and Blue Origin with their self-landing boosters. Not far behind is Rocket Lab, delivering its first commercial payload in November of last year. Then there are groups taking new approaches to putting cargo in orbit. Relativity Space (relativityspace.com) is leveraging 3D printing to build its rockets much faster than current manufacturing can — and therefore, allowing for much cheaper payload deliveries. To do this, they've had to build the world's largest metal-based 3D printer, dubbed Stargate. Currently doing ground-based tests, they aim to do their first launch in 2020 and their first commercial flight in 2021. From there, the idea is to go interplanetary. "Where we really see 3D printing going, in our long-term vision, is to 3D print the first rocket made on Mars," says co-founder Tim Ellis.

VIRGIN ORBIT

Another alternative rocket concept comes from an offshoot of Virgin Galactic. Called Virgin Orbit (virginorbit.com), the company's mission is to help put the new style of small satellites, like Planet's CubeSats, in place. Realizing that fixed launch sites limit possible rocket trajectories and are subject to weather delays, they designed LauncherOne, a 70'-tall, 16,000-pound, two-stage rocket that straps under the wing of a 747. The plane flies to high altitude, then drops the rocket far above the ground. A few seconds later, and a great distance away, it ignites and blasts into space. "In many ways those first couple tens of thousand feet are the hardest part of space flight, because that's where the atmosphere is the thickest," says Will Pomerantz, Virgin

to support the solar panels, and used guitar string for the antenna wires."

Totally solar-powered, and with the ability to transmit over very long distances, the lightweight systems use an ordinary flotation method. "I use a very inexpensive 36-inch silver foil party balloon made by Qualatex for best success, which will float around 29,000 feet and will stay aloft upwards of three weeks," Brown says. "Students learn about weather patterns, buoyancy and lift calculations and experience the thrill of operating a mini-space program as their payload flies around the world. They can track their experiment in real-time with their smartphones on tracking maps which display the location, altitude, solar panel voltage, and temperature."

COMMERCIAL

While hobbyists and students find new ways to do research in and near space, engineers are also building new commercial endeavors, getting incredible programs off our planet and creating the designs that will allow humans to go further than ever.

PLANET

One of the most interesting new space business successes comes in the form of an array of 120 camera-equipped satellites that circle the Earth every day to provide daily updated images of the planet's entire landmass. The company behind it, Planet (planet.com), began in San Francisco in 2011. The foundation of the business is leveraging small, cheap electronics to make satellites that are essentially disposable; their short lifespans allow for frequent replacements that feature newly updated technology, as opposed to the classic NASA model of building large, long-life satellites that are only upgraded through tricky space rendezvous. "The design gets updated roughly every three to six months," says Mike Safyan, Planet's VP of launch, of their primary CubeSat Dove satellite series. "Whenever there's an improvement in battery technology, or faster processors operating at lower power, or different sensors we can include or improve, that gets folded into the next generation."

Through multiple launches, and a few acquisitions, Planet now controls a network of satellites that give daily aerial updates and can zoom in on areas of interest. This has led to successful partnerships with groups that need that data, including agriculture and government. But it's also been a boon for crisis management, such as with rescuers helping after the 2015 earthquake in Nepal. "By looking at our

Orbit's VP of Special Projects. They too are in the testing phases; the first orbital flight is slated for mid-year 2019.

NEXT-GEN SPACESUITS

With the retirement of the Space Shuttle program in 2011, space enthusiasts have been awaiting NASA's return to putting humans in orbit. Those plans are now underway with Orion, a NASA/Lockheed endeavor, along with SpaceX's Crew Dragon and Boeing's CST-100 Starliner.

Kavya Manyapu is one of the Boeing engineers hard at work on the Starliner, helping develop the flight crew operations and human systems integration. But she's also an aerospace sciences Ph.D. with patents on a high-tech spacesuit material, and she's applying that research to make what may be worn by the Starliner crew as the craft takes flight. With recent announcements of new lunar missions, there's a very good reason for this material focus. "When we went to the moon back in the 1960s and 70s, we didn't realize how bad the impact of dust could be on mission operations, hardware, spacesuits, and the health of the astronauts," Manyapu says. "Now that we have plans of going back, it's important that we address it."

Her carbon nanotube-based material is currently undergoing dust-defeating tests in the laboratory, and will soon be subjected to the rigors of outer space as part of the MISSE-X program, which places samples of materials outside the ISS for extended periods — up to four years. If that goes well, it's possible that suits employing her material will be used for the proposed human moon landing slated between 2028–2030, and on Mars and beyond after that.

• • •

These students, hobbyists, and specialized engineers are creating and leveraging the new opportunities that have emerged in recent years. And with them come amazing new moments, which in turn birth even more excitement.

The ISS National Lab's Barstow sums it up: "Apollo 11 was so inspiring, but now we have a whole new era where we're inspired by what we can do ourselves — participate in the space program, not just watch it — and that's the fundamental shift." ⊘

F Planet's CubeSat Dove satellites (inset) ring the Earth to take photos like this one of Singapore.

G Stargate, the massive rocket-building 3D printer from Relativity Space, and an 11-foot fuel tank it made in three days.

H LauncherOne, Virgin Orbit's small-satellite rocket, moves into place under the wing of its 747 platform.

I Kavya Manyapu in Boeing's Starliner suit (right); her carbon nanotube-embedded smart material repels and removes gray lunar dust simulant (left).

Fly Like
SpaceX

BUILD YOUR OWN THRUST-VECTORED MODEL ROCKETS AND LEARN THE SAME TRICKS AS ELON — EVEN VERTICAL LANDINGS!
Written by Joe Barnard with Keith Hammond

IF YOU LOVE ROCKETS, YOU CAN'T HELP BUT NOTICE THAT REAL SPACE LAUNCH VEHICLES LIFT OFF THE PAD SLOWLY, but model rockets zip up like darts. That's how I became obsessed with using thrust vector control (TVC) — gimbaling the rocket motor — instead of fins to keep model rockets upright, so they can launch, and land, far more realistically.

ZERO TO ROCKETEER
I started out from scratch in rocketry; I'm all self-taught. After graduating with a degree in audio engineering from Berklee College of Music in 2014, I saw what SpaceX and other aerospace companies were going for with propulsive landing technology and I was hooked. I knew I wanted to get into rocketry to get a job at one of these companies, and I wasn't in a position to pay for another college degree. I figured instead I could demonstrate what I was teaching myself by propulsively landing a model rocket the same way SpaceX landed the Falcon 9. It was a literal "shower idea."

I started BPS in 2015 with the goal of achieving vertical takeoff and vertical landing (VTVL) of a scale model Falcon 9 rocket. This would require me to solve two tough problems — thrust-vectored flight and propulsive landings — using solid-fuel hobby rocket motors.

I picked up a few textbooks (I strongly recommend *Rocket Propulsion Elements* by George Sutton and *Structures* by J.E. Gordon), found a few good YouTube tutorials for coding and mechanical design, and got to work experimenting.

I naïvely thought it would take four months. My first ten launches were failures. But the eleventh succeeded, and the successes accelerated after that. After four years of hard work, crashes, and iterative designs (Figure A on the following page), I'm now achieving beautiful thrust-vectored launches of several rockets, including my 1:48-scale Falcon Heavy — three cores! — that you can watch on my YouTube channel.

And after some very near misses, I'm confident my Echo rocket will stick the vertical landing in 2019. Rather than attempt to throttle a solid-fuel motor, I'm firing the entire retro motor —also TVC'ed —at the precise time and altitude to enable a soft touchdown. It hasn't been easy.

KITS FOR MAKERS
Of course this technology is still not mature, and it's my hope that the advanced model rocketry community will build upon what I've done. In 2017, I began selling my Signal flight computer board and TVC motor mount together in a kit. After getting user feedback, the computer was redesigned from the ground up to include Bluetooth — Mission Control from your phone!

BPS.space is now a proper company and a full-time job for me, funded through flight computer sales, the BPS.space Patreon page, YouTube ad revenue, and sponsorships.

This kit is for advanced rocketeers. If you don't have experience with scratch-built rockets, I recommend you hone your skills first with an Estes ready-to-fly kit and seek advice from fellow rocketeers at the National Association of Rocketry Facebook Group and the Rocketry Subreddit. And if you've got some experience, I hope you'll give it a try! I've even shared a scale model of the Rocket Lab Electron you can build using my tutorials.

TIME REQUIRED:
A Few Weekends

DIFFICULTY:
Advanced

COST:
$450 and up

MATERIALS
» **Signal R2 thrust vectoring kit for low and mid-power model rockets,** $350 from bps.space/signal. Sales of Signal R2 are limited to U.S. citizens and residents. Kit includes:
• **Signal R2 flight computer** with mounting brackets for 74mm airframe
• **Thrust vectoring motor mount, 74mm, for 29mm motors** includes two 9g servos
• **TVC servo extension cables, 50cm (2)**
• **M3.5 mounting screws**
• **Building and flying instructions** with cutout/drill guides
• **BPS stickers!**

NOT IN KIT:
» **MicroSD memory card** such as Amazon #B004ZICNDA, required for programming and operation of the flight computer
» **Rocket motors, 29mm** I use mostly Estes F15 black powder and AeroTech G8 APCP (ammonium perchlorate composite propellant) motors.
» **Rocket body tube, thin wall, 74mm diameter, 45cm lengths (2 or more)** All parts are built to fit in 74mm airframes, Apogee Components # 10202, apogeerockets.com.
» **Motor mount tubes, 29mm diameter, 30cm lengths (2 or more)** One to line the outside of the motor, the other to align the TVC mount. Apogee #10111 works well.
» **Batteries, 9V alkaline or 11.1V LiPo** such as Amazon #B01N32628C
» **Battery connector** with exposed-end wires

TOOLS
» **Hobby knife**
» **Epoxy/glue**
» **Screwdrivers**

JOE BARNARD runs a project and business called BPS. space. He builds amateur rocketry components, primarily electronics with a few mechanical components, to make model rockets look and work like their real-life counterparts. His goal is to keep developing model rockets that match the pace of advancement in the space-launch industry.

A BPS has produced a lot of rocketry flight computers. Like maybe way too many. Here they are in chronological order.

B

C

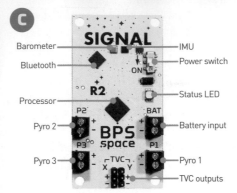

Barometer — — IMU
Bluetooth — — Power switch
— ON
Processor — — Status LED
P2 — BAT
Pyro 2 — — Battery input
Pyro 3 — — Pyro 1
TVC X Y
— TVC outputs

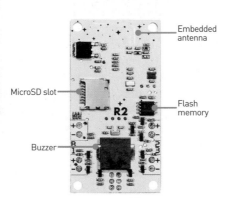

Embedded antenna
MicroSD slot —
— Flash memory
Buzzer —

D

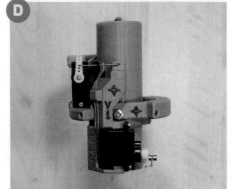

E

THRUST VECTOR CONTROL
Ready to Aim Fire

Model rockets have fins and launch quickly, but real space launch vehicles don't; they actively aim — *vector* — their rocket exhaust to steer the rocket (Figure **B**). With thrust vectoring, your model rockets can slowly ascend and build speed like the real thing, instead of leaving your sight in seconds.

THE COMPUTER

The Signal R2 flight computer (Figure **C**) runs a high-speed control loop, prioritizing functions depending on the progress of the flight. Thrust vectoring draws considerable current, so once burnout is detected, Signal centers and locks the vectoring mount. Focus is then set on detecting apogee and triggering pyro events. It needs at least 8V; 9V alkalines or 11.1V LiPos recommended.

THE TVC MOUNT

Developed over three years of iterative design, the thrust vector control motor mount is made from 3D printed PLA material (Figure **D**). It uses two 9g servos, geared down for higher accuracy. The assembly can gimbal a motor ±5 degrees on each axis, X and Y (Figure **E**). Though the mount will work with up to 40N of force, it works best with lower impulse motors, especially those with long burn times.

THE SOFTWARE

The flight software tracks vehicle flight dynamics while the rocket is powered on. Signal looks for cues to shift system states at liftoff, burnout, apogee, and landing. This makes operation simple — once the settings file is configured for flight, just turn on the flight computer and it automatically enters Pad-idle mode (Figure **F**). Signal can detect launch in under 10ms. Once

detected, thrust vectoring is activated, in-flight abort is armed, and high-frequency data logging begins.

THE DATA

In-flight data logging takes place at 40Hz. Vectoring output, vehicle orientation, altitude, velocity, acceleration, and several other data points are recorded using a custom protocol to a high-speed flash chip. Upon landing detection, Signal creates a new CSV file on the microSD card, dumping flight data into it. Once the data is verified to match, the flash chip is cleared and Signal is ready to fly again. A 1GB card can store hundreds of flights. Flight settings are programmable in a settings file on the card.

THE APP

Signal is configured using an app on an iOS or Android smartphone (Figure **G**). The app helps the user configure TVC sensitivity,

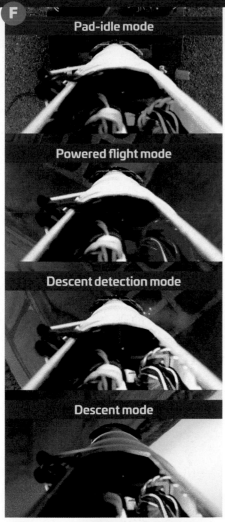

Pad-idle mode

Powered flight mode

Descent detection mode

Descent mode

Pyro Altitudes (meters)		
Pyro 1:	DISC	20.0
Pyro 2:	DISC	0.0
Pyro 3:	DISC	0.0
On-Time (sec):		0.0
Flight Abort:		
Abort at (deg):		20.0
Abort Disable Time:		0.0
	Pyro Testing	

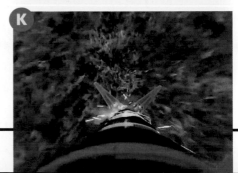

parachute deployment altitudes, the abort system, ground testing, rocket tuning, and more. My goal is to put Mission Control in your pocket.

Build your own scale model Rocket Lab Electron (Figure **H**) by following the "Build Signal R2" tutorials at the BPS.space YouTube channel, and try it out!

PROPULSIVE VERTICAL LANDING
Drop It While It's Hot

When I started building my Signal TVC kits in spring 2017, I put my propulsive landing project on hold. This past year I got back to it: I started drop-testing my experimental Echo rocket from a drone and began a new YouTube series, "Landing Model Rockets."

The series explains my entire process, from selecting motors, to planning the flight profile, to developing a new control board for propulsive landing. It comes in two flavors: Blip, a DIY version using breakout boards and through-hole components; and Blop, a lighter, surface-mount PCB. Both are based on the powerful MK20DX256VLH7 processor that's used in the Teensy 3.2 microcontroller, with a Bosch BMP280 barometric pressure sensor and InvenSense MPU-6050 inertial measurement unit (IMU). For now I'm sharing these experimental boards with my Patreon supporters.

So how can you land with solid rocket motors if you can't throttle them? It's all about timing! If the flight profile is fairly well known and the legs are built to withstand small hops and drops, the motor can be fired at just the right time to slow the vehicle down for a soft landing. As the rocket nears the ground, the microcontroller reads the barometric altimeter and fires the retro motor (Figure **I**). Four landing legs deploy by rubber band (Figure **J**), and the rocket lands upright (Figure **K**). In theory.

Turns out vertical landing is super hard! As the rocket slows to a near-hover, horizontal drift becomes an issue. I'm already deep in the weeds programming the math necessary to control for this. I'm also experimenting with a tiny LIDAR sensor for really precise rangefinding to the ground.

After several very near misses (watch them on YouTube), I refuse to not stick the landing in 2019! Follow me on Twitter to stay up to date.

MY LITTLE FALCON HEAVY
MEET THE MOST ADVANCED SCALE ROCKET ON EARTH.

Myself, the Falcon Heavy model, and Tim Dodd, The Everyday Astronaut.

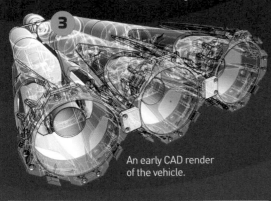

Falcon Heavy model — Flight 1.

An early CAD render of the vehicle.

In December 2017, I set out to build a 1/48 scale model of the SpaceX Falcon Heavy rocket (Figure **1**) as a technology demonstrator for BPS.space, to showcase the advancement that's now possible at the model scale. As usual, I underestimated it as a four-month job! I finally launched it in November of last year (Figure **2**).

Each of the three cores (Figure **3**) carries a flight computer, parachute deployment system, and TVC assembly. The upper stage flies with thrust vector control as well, and it carries a 3D-printed sports car — no Falcon Heavy model would be complete without one.

AVIONICS
The Falcon Heavy flight computers are upgraded versions of Signal (*see page 32*). Each computer has MEMS gyroscopes and accelerometers for sensing movement and orientation on the rocket, the same kind of sensors found in most smartphones. A barometric pressure sensor determines the rocket's altitude above the ground.

On each flight computer, a 48MHz Cortex M0 processor reads the flight sensors at 400Hz and logs 31 channels of data to a flash chip, 40 times per second. Every second that Falcon Heavy is in the air, 4,960 points of data are recorded onboard — never too much data.

The flight computers don't communicate with each other during flight; the sensors and software are accurate enough that it's not required for short flights. None of the flight settings are hard-coded; depending on the flight profile, they can all be changed using the Signal iOS/Android app.

The flight software is written in C++, the iOS app in Swift, and the Android app in Java. Flight guidance is computed using *quaternions*, a complex number system for 3D orientation and rotation that's more computationally efficient than other methods (see youtu.be/d4EgbgTm0Bg for a wonderful introduction).

MECHANICS
The attachment points at the top of each side core slide down a ramp on the center core, giving them a bit of clearance during stage separation (Figure **4**). This passive setup keeps things very simple during flight, but the cores are usually bolted together while the vehicle is on the ground. All three cores are also connected at the base of the vehicle using a slightly simpler thrust plate.

The side cores remain attached to the vehicle by maintaining a slightly higher net thrust force than the center core/upper stage. As soon as the center core produces more net

thrust than the side cores, it will pull away and the stages will separate. For flexibility, each side core also has a slot for a small separation motor; I haven't needed to use sep motors on flights so far, but they may be helpful later down the road.

The center core of the Falcon Heavy model goes through two boost phases during flight. During the first phase, the side cores are attached. Right around side core burnout, the center core lights a second motor, and the second boost phase begins. These two rocket motors are mounted on top of each other in the center core's thrust vectoring mount. When the second motor ignites, the lower, spent motor, is ejected. This same technique, called *hot staging*, is used to control ascent and propulsive landing motors in other BPS.space rockets.

The rocket motors in each stage of the Falcon Heavy model can be gimbaled ±5 degrees in any direction. Because the side cores are not firing directly through the vehicle's center of mass, they can be used not just for pitch and yaw, but *roll control* as well. Each multi-core flight runs a roll program which usually targets 20 degrees of positive roll, executed at 30–40 degrees per second.

Launching the Falcon Heavy model requires a bit of forethought. The center core motor has a slight thrust spike at ignition, after which the amount of force produced slowly tapers off. At liftoff, the side cores must have a greater net thrust than the center core. Because of this, the center core is lit at T−1 second while the vehicle is still held on the pad by the launch clamps. At T−0 the side cores are lit, and at T+0.25 the beast is released.

Also, do I even have to say it? Of course those boosters are gonna land! The propulsive landing test program is isolated from the Falcon Heavy program right now, but the two will merge as the success rate for both programs increases.

LAUNCH PAD 2.0
After five iterations, Launch Pad 2.0 was redesigned from the ground up to support both single- and three-core rockets. (You can see it in Figure H on page 33.) With 8 total launch clamps, and an easily modifiable iron flame trench, Pad 2.0 is my most flexible platform yet. My Impulse launch computer, based on an Atmel ATSAMD21 microcontroller, has plenty of inputs and outputs to support add-ons like load cells, wireless communications, and any other peripheral that communicates through I^2C or SPI.

WHAT'S NEXT?

» **Fine-tuning the TVC** — I continue to launch the Signal R2 board on my new Scout D1 rocket to improve the tuning feature in the software.

» **Slow-burning propellant** — Most model rocket motors burn just 3–7 seconds, not long enough to simulate real space vehicle launches. I've been working with Aerotech to develop long-burning motors such as their new G11 and G8ST that burns for almost 18 seconds!

» **High-powered rockets** — I've been getting involved in the HPR community, where amateur rocket flights go miles high and safety is a lot more important. So I'm now developing the Arc dual deploy board: 2 pyro channels for firing main and drogue chutes, barometric altimeter up to 100,000 feet, multi-speed data logging, and Bluetooth for smartphone configuration (Figure **L**).

» **Reaction control system (RCS)** — One more thing: Rather than attempt TVC on high-power motors, I'm developing a reaction control system — six little cold-gas thrusters near the top of the rocket to improve stability (Figure **M**) — and of course a new Relay RCS board (Figure **N**) to run it. ◎

[+] MORE FROM JOE AND BPS.SPACE:
Twitter: twitter.com/joebarnard
YouTube: youtube.com/channel/UCILl8ozWuxnFYXIe2svjHhg
Patreon: patreon.com/bps_space

MOTOR MOUTH
YOUR ROCKET MOTOR QUESTIONS ANSWERED

Q: Do you make your own motors?
A: Nope, I use commercially available propellants. Usually black powder motors from **Estes** or APCP motors from **Aerotech**. For custom or more complicated builds, BPS.space outsources propellant work to other manufacturers, who can achieve higher precision and predictability. I learn best by experimentation, which works very well for software and electronics, less so for explosives and propellants.

I also got a C+ in chemistry during high school — designing and making rocket propellant is not a good idea for me. :)

Q: I want to make my own rocket motors. Do you have any advice?
A: I strongly **recommend against** building your own rocket motors unless you have an experienced mentor or teacher to help you. There's certainly information about building these motors online, but much of it is incorrect or lacks enough consideration for safety.

Q: I heard that active control in model rockets is illegal, can't you get in trouble for that?
A: Nope, it's not — **active control** has been around in model rocketry since the 1980s! None of the U.S. regulations prohibit active stability in model rockets. There's an important distinction here between guidance and stability.

Guidance is usually about maneuvering in reference to one or more real points in space. This might be via GPS, GLONASS, RF, or even dead reckoning.

Stability is just about keeping the rocket upright, which serves the same purpose as fins on a traditional model rocket. Signal does not carry any GPS or point-based guidance equipment. The computer is only able to keep the rocket vertically stable off the launch pad.

Written by ● Craig Couden

Imagination Exploration

BRING THE COSMOS CLOSER WITH THESE SPACE-THEMED PROJECTS

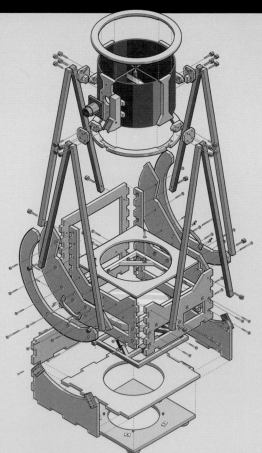

BACKYARD DOBSONIAN TELESCOPE

Dobsonian telescopes are popular with amateur telescope makers for their ease of design and construction, portability, and their use of large optical mirrors. This one combines a Newtonian reflector telescope with a unique two-axis movable base, using a primary mirror to capture and reflect light, a secondary mirror to direct light into an eyepiece, and a focuser to make fine adjustments for viewing. And if you really want to get serious, you can grind your own lenses for about 40 hours a pop. makezine.com/go/dobsonian

SPACESHIP BUNK BED

Pete Dearing was undoubtedly in the running for maker dad of the year for the interactive cockpit in these spaceship bunk beds. With a build time clocking in at over 100 hours, most of the buttons control real lights, sounds, fans, meters, and headlights via Raspberry Pi. You can find the base plans for the bed online as a starting point for your own starship build. makezine.com/go/space-bunk-beds

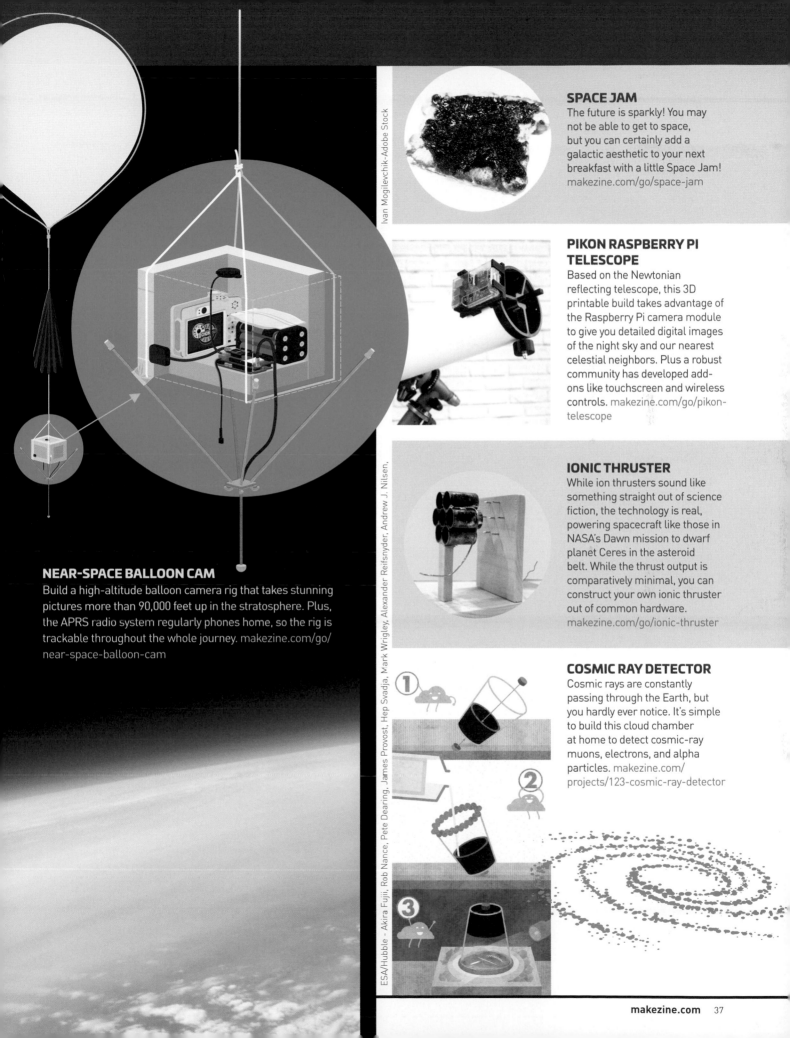

SPACE JAM
The future is sparkly! You may not be able to get to space, but you can certainly add a galactic aesthetic to your next breakfast with a little Space Jam! makezine.com/go/space-jam

PIKON RASPBERRY PI TELESCOPE
Based on the Newtonian reflecting telescope, this 3D printable build takes advantage of the Raspberry Pi camera module to give you detailed digital images of the night sky and our nearest celestial neighbors. Plus a robust community has developed add-ons like touchscreen and wireless controls. makezine.com/go/pikon-telescope

IONIC THRUSTER
While ion thrusters sound like something straight out of science fiction, the technology is real, powering spacecraft like those in NASA's Dawn mission to dwarf planet Ceres in the asteroid belt. While the thrust output is comparatively minimal, you can construct your own ionic thruster out of common hardware. makezine.com/go/ionic-thruster

COSMIC RAY DETECTOR
Cosmic rays are constantly passing through the Earth, but you hardly ever notice. It's simple to build this cloud chamber at home to detect cosmic-ray muons, electrons, and alpha particles. makezine.com/projects/123-cosmic-ray-detector

NEAR-SPACE BALLOON CAM
Build a high-altitude balloon camera rig that takes stunning pictures more than 90,000 feet up in the stratosphere. Plus, the APRS radio system regularly phones home, so the rig is trackable throughout the whole journey. makezine.com/go/near-space-balloon-cam

Lunar Mission

NO POWER TOOLS NECESSARY TO MAKE THIS IMPRESSIVE, GIANT BACKLIT MOON

Written and photographed by Caleb Kraft

TIME REQUIRED:
2 Hours

DIFFICULTY:
Easy

COST:
Less than $100

MATERIALS

- » **Shower curtain with printed full moon** I used the Emvency Luna Full Moon curtain, Amazon #B07D8MSDM8, but there are lots of choices out there.
- » **Frosted shower curtain** the same size as the full moon curtain, to act as a diffuser
- » **¾" PEX tubing, about 20'**
- » **¾" PEX tube mounting clips (8)**
- » **LED strip, RGB or solid color, with remote**

TOOLS

- » **Hacksaw**
- » **Scissors**
- » **Staple gun**

CALEB KRAFT has always been a huge fan of space and has found a way to bring a bit of that into his home.

I LOVE THE MOON. LIKE MOST FOLKS, I TAKE IT FOR GRANTED MOST OF THE TIME, but occasionally I'm reminded that it's a giant freaking rock orbiting us about 240,000 miles away. I'm not sure why, but I find that fascinating.

I wanted to bring a bit of the beauty of the moon into my home, but like usual for me, I had to make it into some kind of big impressive item.

THE IDEA

I knew I wanted my giant moon in high resolution and backlit, and I didn't want it to take up a bunch of room, so it had to be pretty flat. I've got really high ceilings in my living room, so I knew I could go huge without it feeling weird and claustrophobic.

My initial thoughts jumped right to the giant CNC router I have access to, or 3D printing something in multiple parts and assembling it into a massive structure.

But after some consideration, I ultimately decided to challenge myself to do it low-tech, and I wanted to use common tools that most people have within reach. In the end, you could pull off this whole thing with only a handsaw, a staple gun, and a pair of scissors.

THE PRINT

I hunted for high-resolution images of the moon and found plenty. Most required ordering a custom print. After a few calls to sign shops it turned out that I was going to be paying over $200 for a 6-foot-wide, high resolution print of the moon on plastic. There had to be a better way.

I considered projecting an image of the moon onto a vinyl shower curtain, painting the craters, then stretching that over a frame. During this brainstorming session, it hit me ... I wonder if there are already shower curtains out there with high-resolution images of the moon on them? Well, sort of.

I found two shower curtains (Figure Ⓐ). The larger one was roughly $30, while the smaller one was $40. I decided to use the larger just for the impressive size alone. The resolution isn't as high as I'd like, but from a few feet away, you'd never know. The image is also somewhat elongated and overlaps with the grommets at the top.

I also purchased some cheap frosted vinyl shower curtains to act as diffusers. You can see the curtain, diffuser, and frame all laid out in Figure Ⓑ.

THE HOOP

For the frame, I wanted to use something that you could construct with few or no tools. I remembered that people make hula-hoops out of PEX tubing all the time, so I went that route. You can saw through the tubing pretty easily. I used a pull-saw meant for wood, but you could probably get through it with a steak knife if you were persistent enough.

Another great thing about PEX is that it is soft enough to staple into. I simply stretched the shower curtains over the frame and stapled them in place (Figure Ⓒ).

After that, cut off the excess shower curtain to clean it up.

THE LEDS

I used an RGB LED strip. I chose a cheap one, and the white setting is super blue (Figure Ⓓ). Doing it over I probably would look for a truer white strip and skip RGB all together.

Applying the LED strip was easy enough. It has an adhesive backing, and sticks well to the PEX tubing (Figure Ⓔ).

MOONRISE

Originally, I wanted to just hang it all from a single hook. But due to uneven stresses on the somewhat flexible frame, the whole thing was slightly Pringle shaped. This really wasn't a huge problem though: They sell clips for PEX that you can simply press the pipe into and it grips it pretty well. I actually chose to cut some of the tip off of the clip to make the grip a bit less extreme.

Once I had a ring of these screwed into my wall, all that was left was to press the moon into place! You can see how incredibly thin this thing is with the clips (Figure Ⓕ).

GOING FURTHER

There are so many ways to improve this that popped into my mind as I was building it:

1. Improve the light color
This is easy enough to fix. Just order better LEDs.

2. Improve the existing light diffusion
This one has me a bit stumped. To get good diffusion and light up the entire moon, I would have to pull it out much further, but I don't want it that far from the wall! How could I diffuse the existing LEDs better to the center?

3. Add more light
Similar to number 2, how could I better diffuse lights in the center without making it fatter? I suspect I'd get hot spots even with very aggressive diffusion with the current thin design. A different light may be needed.

4. Add a microcontroller to make the lights do something smart (like display moon phases)
You could totally plop an RTC onto an Arduino and light up a few light strips with various moon phases. That was initially part of the plan. While it does sound neat to be able to do that, I found myself wondering why in the world you'd ever light up a partial moon phase when you could have a full moon all month long? ⊘

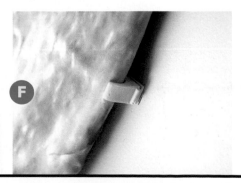

[+] Watch the video at makezine.com/go/illuminated-wall-moon

Where in the World Is the ISS??

THREE WAYS TO LOCATE HUMANITY'S ONLY SPACE OUTPOST AS IT HURTLES OVERHEAD
Written by Keith Hammond

KEITH HAMMOND built a 36" Revell model Apollo Saturn V rocket with his dad in the 3rd grade and took it to school, where it fell to Earth and shattered into a million pieces. His launch safety record since that day is spotless.

WHEN CHINA'S TIANGONG-1 BURNED UP LAST YEAR, IT LEFT THE INTERNATIONAL SPACE STATION (ISS) as humanity's sole outpost in space. Today six astronauts live aboard the ISS, performing spacewalks and science experiments, filming scenes for an upcoming virtual-reality series, and unloading cargo deliveries from capsules like Northrop's Cygnus and SpaceX's Dragon.

These brave astronauts orbit the Earth 15 times daily at 17,000mph, passing over your head 5–8 times each day. It's fun to try to spot the ISS at night, and you can take photos of it streaking across the dark sky.

So how can you find the ISS? Sure, there are websites (spotthestation.nasa.gov) and apps (issdetector.com) that'll tell you when and where to look. But you're a maker. You want to build something cool!

DESKTOP ISS POINTERS

The ISS pointer is a little robot arm that literally points to where the station is in space at any given time. It's a challenge that makers approach in many ways, but the bare necessities are a microcontroller or mini computer, a stepper motor to control azimuth (horizontal motion), a servo or stepper to control elevation (altitude), and of course computer code that can figure out where to point.

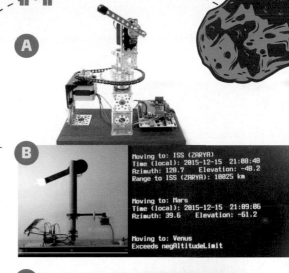

Grady Hillhouse's Desktop ISS Orbit Tracker (Figure **A**) is a robust prototype built with Actobotics parts, chain drive, and an STM32 Nucleo microcontroller board. This build lacks Wi-Fi access, so the ISS orbits are hard-coded periodically into a C++ version of the standard SGP4 satellite tracking software. Build and code shared at www.instructables.com/id/International-Space-Station-TrackerPointer and github.com/gradyh/ISS-Tracking-Pointer.

Patrick Ferrell's adaptation, The Thing Pointer (Figure **B**), uses a Raspberry Pi, belt drive, light-up 3D printed finger, and LCD display, and adds an Adafruit GPS Hat and 3-axis magnetometer so the bot always knows where it's at and where it's pointed: at the ISS, the planets, the moon, and more. The Python code uses the PyEphem package (rhodesmill.org/pyephem, built on the SGP4 algorithms) to track the position of astronomical objects and satellites, and updates the orbital parameters from the web every couple of days. See it at youtube.com/watch?v=9uxs_4G7Sfo.

Developer and pentester K4YT3X built a pointer with a RaspPi and 3D-printed gears, and open sourced the code at github.com/K4YT3X/iss-pointer; it also uses PyEphem.

Andrey (Destroyer2012) designed a totally 3D printed pointer mechanism with really cool gearing (Figure **C**). He's running it with an Arduino Nano, real-time-clock module, and optical endstops. It's still a work in progress at thingiverse.com/thing:2043674.

The simplest mechanical build I've found is Russell Grokett's ISS Pointer (Figure **D**) using the Adafruit Huzzah ESP8266 board — he mounts the altitude servo right on the azimuth stepper's shaft. He runs it from a Raspberry Pi (PyEphem again), and adds a 16×2 LCD display and mini speaker for sound. Build and code shared at www.instructables.com/id/Building-an-ISS-Pointer-Tracker-Using-Adafruit-HUZ and github.com/rgrokett/ESP8266_ISSPointer.

But by far the cutest is Don Core's ISS Pointer Robot (Figure **E**), using Grokett's code. Prototyped with spare parts and Lego, it's got an OLED display, an old hard-drive arm for a pointer, and incandescent light bulb eyes that shine when the ISS rises high enough to see. Build at www.instructables.com/id/ISS-Pointer-Robo.

If you're starting your own ISS pointer project, note that PyEphem author Brandon Rhodes is replacing it with a more powerful library called Skyfield; learn more at rhodesmill.org/skyfield.

ISS-ABOVE HD VIDEO DISPLAY
issabove.com

If you've got room for something bigger and splashier, ISS-Above is a fantastic project for the classroom, bedroom, or makerspace — a dedicated video tracker display that also provides astronaut's-eye views from the ISS. The project was launched on Kickstarter by Liam Kennedy (see profile on page 27) and there are now 3,000 of these in the wild.

A Raspberry Pi calculates the space station's position in real time and sends HD video to a TV or monitor showing when the ISS will next be present in your sky (Figure **F**). When the station approaches, a PiGlow LED display starts flashing and the info screen tells you where to look. When it's overhead, the screen switches to live views of Earth from ISS ("I can see your house from here!"), maybe a Soyuz or Dragon docking, and whatever else is captured by ISS' external cameras. It's awesome.

Buy the basic setup for $148 (RaspPi, PiGlow, transparent case, power supply, and SD card with ISS-Above software) and provide your own monitor and HDMI cable. Or DIY it and just grab the SD card with software for $42 if you've already got a Pi to run it on; it's compatible with a number of LED and LCD displays like ThingM Blink, Adafruit LCDs, and the Kano 10-LED ring. There's an optional 3D printed bracket to mount the ISS-Above on your TV, and a teacher's handbook for schools.

ISS NOTIFIERS

This is an easier type of project that simply lights up or raises a flag when the ISS is in your sky — you can consult websites or apps to figure out where to look, but it's nice just to be reminded when those astronauts are overhead.

Cat Haines built a sharp-looking, light-up, laser-cut display (Figure **G**) using the Electric Imp IoT platform, Adafruit Neopixels, and data from the website wheretheiss.at, and shared it at www.instructables.com/id/ISS-Overhead.

And for a fun beginner's build, check out TokyLabs' pop-up ISS Notifier project on the next page. ●

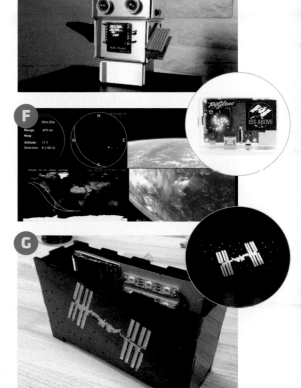

NASA, Grady Hillhouse, Patrick Ferrell, Destroyer2012, Russell Grokett, Don Core, ISS-Above, Cat Haines

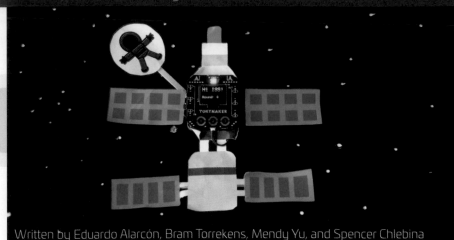

Written by Eduardo Alarcón, Bram Torrekens, Mendy Yu, and Spencer Chlebina

Build an Easy **ISS Notifier**

FLY AN ASTRONAUT FLAG EVERY TIME THE SPACE STATION PASSES OVER YOUR HOME

YOU CAN EASILY CREATE A HARDWARE NOTIFIER THAT LIFTS UP A PAPER ASTRONAUT TO ALERT YOU each time the International Space Station passes over your location. Way more fun than a text!

Tokymaker is a microcomputer from TokyLabs that lets you create inventions in 5 minutes by mixing electronics, programming, and IoT — with no prior engineering knowledge. Electronic modules connect without soldering, and everything is open source. It's programmed from a website, which sends code over Wi-Fi — no cables, software, or plugins. Using the graphical language Google Blockly, even non-programmers can easily create code.

1. SET UP ADAFRUIT IO FEED

Create a cloud account at io.adafruit.com. Then click on Feeds→Actions→Create a New Feed. Name it "ISS." Click the View AIO Key button, then copy your unique *key* somewhere safe — you'll need it later to link your Tokymaker to your Adafruit IO feed.

2. SET UP IFTTT ACTION

Create an account at ifttt.com. This site links Internet services in a very simple way. In our case: *If* the ISS passes over a specific address, *then* send the number 100 to your Adafruit ISS feed.

First you'll choose the *trigger*. Select New Applet, then click on "+ this" and type "Space" in the search bar. Click the Space icon, then choose "ISS passes over

a specific location," then type your address and click on "Create trigger" (Figure **A**).

Next, create the *action*: sending the number 100 to the Adafruit IO feed. Click on "+ that" and choose Adafruit. Click the Connect button and complete the fields in the popup window (Figure **B**). Then, click on "Create action." The cloud setup is done!

3. PROGRAM THE TOKYMAKER

Now for the physical part: Every time the number 100 is in the Adafruit IO feed, your Tokymaker will run a program to turn on a light, move a motor, whatever you want. Go to tokylabs.com/ISS and download the basic ISS Notifier code to your Tokymaker. (Or make it yourself at create.tokylabs.com!)

4. BUILD YOUR ISS NOTIFIER

Cut out the image of the space station, and tape the Tokymaker onto the front. Glue the battery pack on the back. Plug the servomotor into Output 1, wrap its cable around, and glue the servo on the back so it's standing up. Tape the printed astronaut to one end of the stick. Cut the stick to size, then glue the other end to the servo arm so the astronaut faces the front (Figure **C**).

SPACEWALK!

Now whenever the ISS passes over your location, your Tokymaker will move the servo to raise the astronaut, light up an LED, and show a message on the OLED screen with the number of orbits that day! ✪

TOKYLABS, based in Shanghai and Barcelona, envisions technology not as a final objective but as a tool to achieve creative goals.

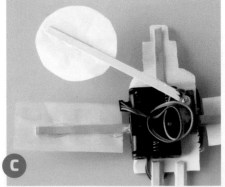

Holden Johnson

[+] For more photos, and tips on the cloud and the code, visit makezine.com/go/tokymaker-iss-notifier.

Galactic Gears

Written by Caleb Kraft

A MULTITUDE OF MATERIALS LEND THEMSELVES TO CREATING THESE MECHANICAL ORRERIES

You've probably seen an orrery before, though you may not recognize the name. An orrery, or mechanical model of the solar system, or sometimes just the Earth, moon, and sun, is a common image in popular culture to show that *sciency* things are happening. Remember the giant mechanical thing from Aughra's lair in *The Dark Crystal*? That's an orrery.

They may seem unfathomably complex, but luckily people have figured out all the gear ratios and published easily reproducible versions on the internet. Pick what material you want to work with, and someone has you covered.

1. METAL
Do you have the chops to grind and polish your own brass mechanisms? This might be the one you're after. This blog details all aspects of creating a truly gorgeous orrery with materials such as brass, aluminum, and glass. zeamon.com/wordpress/?page_id=468

2. 3D PRINTING
Fire up your printer — this solar system model by WidgetWizard on Instructables is fully geared for 8 planets plus the moon orbiting the Earth. While this is mostly 3D printed, there is a considerable amount of work that needs to be done to fully assemble it. www.instructables.com/id/8-Planet-Motorized-Orrery-3D-Printed

3. PAPER/FOAM
Oh yes, we remembered you folks out there who don't have a workshop, fancy tools, or tons of time and money. You can download this template and cut foam and paper to create the "wizard's orrery" in an afternoon. Complex geared systems don't have to be a huge undertaking when wonderful people share detailed plans like this. stormthecastle.com/how-to-make-a/how-to-make-a-wizards-orrery-5.htm

4. LEGO
Jason Allemann has put together a full breakdown of how to build your own sun/moon/Earth orrery from the simple, ubiquitous blocks. It can be either hand cranked or motorized and takes less than 500 parts combining standard Lego and Technic. jkbrickworks.com/earth-moon-and-sun-orrery

5. PLYWOOD
Marbles, plywood, and brass comprise this solar system model. Matthew Strausser fabricated this system using a CNC router, though you could print out the patterns and cut them using a band saw or scroll saw. Though he doesn't include downloadable patterns, he does explain what he based his on, and describes the math needed to create your own, in detail. www.instructables.com/id/Orrery-a-Mechanical-Solar-System-Model-From-Plywoo/

Experimental Thinking

MAGNITUDE.IO WANTS TO MAKE IT EASY FOR KIDS TO ACCESS SCIENCE IN SPACE

TED TAGAMI
is CEO and co-founder of Magnitude.io.

AMBITION, TENACITY, AND A LITTLE MOXIE WILL GET YOU FAR AS A MAKER. SPACE EXPLORATION NEEDS THAT SPIRIT, along with a high level of rigor and discipline. Seeing an opportunity in 2013, I got together with my longtime friend Tony So to launch Magnitude.io to increase accessibility to space by "orders of magnitude," while greatly reducing costs. We did this just after the state of California announced the adoption of the Next Generation Science Standards (NGSS), which would integrate science and engineering practice. Nearly two-thirds of students in the United States are implementing standards influenced by NGSS. At Magnitude, we believe that these new standards are fertile ground for authentic research and exploration in learning environments like public schools.

Magnitude's prime objective is to integrate science and engineering to create authentic learning experiences.

A Lodi Unified high-altitude balloon (HAB) 22 miles above the California coast.

CanSats.

Students from Lodi Unified prepare to release their high-altitude balloon.

Leading with the investigation of science phenomena, we ease into engineering later. This differentiation has helped tremendously as some teachers are confident in physics, chemistry, biology, or mathematics but apprehensive about engineering in a classroom setting. Working with talented people and organizations in their given fields of expertise has allowed us to stay creative and to explore the many exciting ways to engage students through the awe and wonder of space.

A PAIR OF EXPERIMENTS

We currently have two projects running on the Magnitude Learning Platform. One is **CanSat**, a simulated satellite that can be launched in high-powered rockets (HPR) or high-altitude balloons (HAB). The other is **ExoLab**, a network of devices on Earth connected to the International Space Station. Last year we received the 2018 ISS Innovation in Education award for our novel approach to create a network of experiment devices connected to the ISS. Our sixth mission is scheduled to launch aboard the SpaceX-18 resupply mission in July 2019 where we will investigate nitrogen fixing bacteria (*Rhizobium leguminosarum*) in microgravity.

Space isn't simple, and our first missions were not successful. We started by collaborating with a start-up satellite company that was unable to get their

LAST YEAR, MAGNITUDE. IO WAS AWARDED THE 2018 ISS INNOVATION IN EDUCATION AWARD IN RECOGNITION OF OUR NOVEL APPROACH TO CREATE A NETWORK OF EXPERIMENT DEVICES CONNECTED TO THE ISS.

satellites to operate in orbit. After three failed attempts we thought it better to move forward with our own engineering. We wanted to fulfill our promise of providing every student access to an extraordinary experience.

With any maker project, the community is power. Less than a year into the creation of Magnitude, we met Stanford professor emeritus Bob Twiggs, co-inventor of the modern CubeSat, and shortly thereafter were introduced to space systems engineer Twyman Clements, now CEO of Space Tango. With these collaborators, we developed CanSat and ExoLab.

CANSAT

Before CubeSats were invented, Professor Twiggs and his Stanford students built simulated satellites in the volume of a 12-ounce soda can in order to understand the systems required and to significantly

reduce the time, cost, and complexity of satellites. These CanSats were launched in rockets and deployed at apogee. Descending under parachute, their hang time was similar to a horizon-to-horizon pass of a satellite in orbit. CanSats are a global phenomenon; in Europe there's an annual competition for high schoolers (cansat.eu).

The CanSats we've designed are shields built for the Arduino Uno. Breakouts on our custom shield enable us to extend sensor measurements beyond our base sensors (barometric pressure, temperature, humidity, acceleration, magnetic fields, lux, and GPS), to include gas measurements, particulates, even radiation.

We have run dozens of CanSat missions with students across the United States. Some samples:

Atlas Space Explorers is a Scouts BSA organization in Traverse City, Michigan, that is learning about the complexities of engineering real-time communication between the spacecraft and the ground. Working in an FAA-approved flight range, their HAB will fly a CanSat to act as a simulated satellite to track and send commands to a drone beyond line of sight. Their sponsor company is Atlas Space Operations (atlasground.com), a company building the Freedom Ground Network across VHF, UHF, S-Band, and X-Band for missions beyond Earth.

Magnitude.io

ExoLab's model organism, *Arabidopsis thaliana*.

Students from Russell Independent School District, Kentucky, with their ExoLab.

The WEX Foundation is a nonprofit based in San Antonio, Texas. Funded by a NASA Space Education Grant, WEX students are part of a program called LCATS (Lunar Caves Analog Test Site) in which they are researching a new quartz crystal microbalance (QCM) for use on the lunar surface. In its second year with Magnitude, WEX will be flying tests with CanSat and QCM in both HPR and HAB missions later this spring.

Schools across the country use CanSat to investigate the world around them. In Northern California the **school districts of Lodi, Manteca, and Stockton** actively launch HABs and HPRs, while in **San Leandro** students have extended the work with CanSat to build an air quality monitor and have received two Regional Air District grants so far to continue their work!

The CanSat was an inspiration for the modern satellite, the CubeSat. The CubeSat in turn was the inspiration for the cube-sized laboratories like ExoLab which are currently installed aboard the International Space Station. These cube systems are essentially intermodal freight transportation for space. A 40'-long shipping container can be sent anywhere in the world because of standard sizes and equipment to handle them. In space, the equivalent of shipping containers are these 10cm (4") cubes weighing in at 1.33kg (about 3 pounds).

EXOLAB

Our ExoLab project dramatically reduces the cost and complexity of running an experiment in space. Instead of tens of thousands of dollars we are now looking at hundreds of dollars.

Based on the Raspberry Pi, ExoLab is an IoT device that takes a picture of the project every hour and shares ambient information of the experiment underway (CO_2 level, temperature, humidity, light level). This information is connected with other classrooms around the world — all of which are connected to an ExoLab aboard the ISS. The first six missions have all been biological. As we reach out to other educators and groups we will explore adding sensors via the front I2C port.

Some Magnitude.io collaborations:

The Lodi Unified School District in California has enabled their primary school libraries with ExoLab along with a live view of Earth from the ISS (see "ISS-Above" on page 41). Their unique approach encourages reading and writing by tapping into kids' interest in space exploration and science. Librarians curate books and even set up question boxes for young students to ask any questions they have about space. One of my favorites so far has been: "How far can a bunny jump on the moon?" We're excited by the groundbreaking work this district is doing with ExoLab. As of this writing the entire district is contemplating what unique science investigation they would like to run aboard the ISS in 2020!

I-Innovate (i-can-innovate.com) works with students throughout South Africa to explore science, technology, engineering, and math (STEM) skills alongside essential 21st-century skills such as teamwork, problem-solving, critical thinking, creativity, innovation, and computational science.

Some teachers really step up to the challenge. **Michael Wilkinson** is a math and science teacher from the Bronx who has brought his fourth-grade students deeply into the ExoLab experience. He brings so much knowledge and enthusiasm to learning, he has become the principal investigator (PI) for our next mission to the ISS. In addition to this, he is actively developing a cadre of teachers who are passionate about space for upcoming experiments.

Our first partner in Europe is **CodeDoor** (codedoor.org).Founded in 2015 to teach untapped groups employable skills in coding, they now work with global companies to train their employees. This summer we are excited to collaborate with Codedoor and the Media Centers in Central Hesse to bring high schools onto the ExoLab platform from the Mittelhessen region of Germany where they will learn about the platform and develop their own unique experiment for German students. After the initial project we will scale nationwide in Germany.

MOVING FORWARD

Big news: Magnitude.io has just been included in the **Astrobee Guest Scientist** program (see page 26). We are collaborating with gas sensor company Spec Sensors and engineering firm KWJ Engineering, both in Newark, California, to build an electrochemical sensor array — like

Intern Matthew Young readies plant experiments.

Students from Cape Academy prepare their ExoLab.

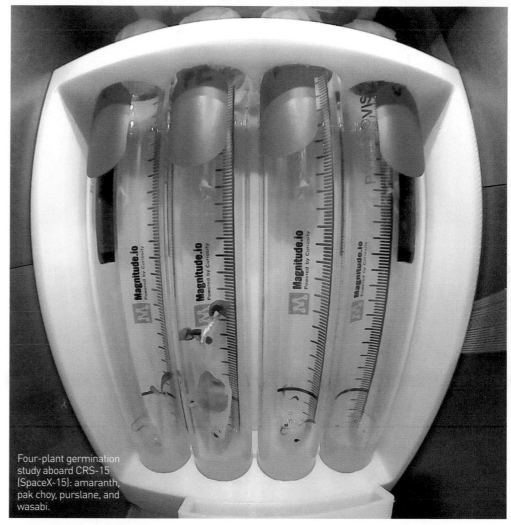
Four-plant germination study aboard CRS-15 (SpaceX-15): amaranth, pak choy, purslane, and wasabi.

a digital nose for the free-flying robot, monitoring the cabin air aboard the ISS. Similar instruments will be available for schools to measure indoor air quality when the project launches in 2020.

Beyond that, we have some interesting projects in development. How about a lunar mission for students within 5 years? Or perhaps a greenhouse on Mars? Being able to tie into link and power budgets on larger missions, we're playfully using a hashtag for the deep space missions: #WeJustNeed1kg.

Getting back to Earth, one amazing

phenomenon that astronauts have expressed is what's called the *overview effect*. This cognitive shift in awareness of seeing our planet hanging in the infinite darkness has made some acutely aware of how precious our world really is. While we are deeply passionate about space exploration, there is still so much to discover here on our beautiful planet. To that point, our web-based platform for learning will be opening up other research and exploration opportunities right here on Earth. Stay tuned! ◐

Magnitude.io

Habitats for Humanity

THE MARTIAN BASES OF THE FUTURE MAY BE MADE FROM REGOLITH AND WATER-ICE

Written by Melodie Yashar

1. Crew quarters
2. Crew unit
3. Hygiene unit
4. Greenhouse
5. Food prep
6. Library
7. Wardroom
8. Soft hatch
9. Airlock
10. Ice chambers
11. Gas insulation pockets
12. Vision window

IN 2015 MY COLLEAGUES AND I FOUNDED SEARCH+ (SPACE EXPLORATION ARCHITECTURE), with a mission to conceive, investigate, and develop innovative "human-centered" designs enabling human beings not only to live, but to thrive in space environments beyond Earth. We are a collective of designers, architects, technologists, and researchers that integrate design thinking with traditional engineering workflows to establish the importance of designing for the human experience and holistic health in long-duration space missions. Our work consists of concepts for human habitats on planetary surfaces, in-transit vehicles, as well as the technologies that enable their deployment. Within these environments we consider not only the appropriate concepts for long-term habitation, research, and working modules in space, but also the human interfaces with hardware and technology that keep the astronauts alive and thriving.

Members of SEArch+ are professors at Pratt Institute's School of Architecture and Design and are the recipients of NASA

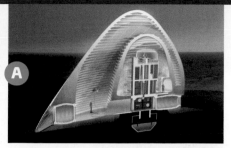

MELODIE YASHAR is an architect, designer-researcher, and co-founder of SEArch+ (Space Exploration Architecture). Her background is in architecture, industrial design, and human-computer interaction.

A Mars Ice House section perspective.

B Impact test from Construction Level 1 of the Phase 3 3D-Printed Habitat Competition.

C Final structure from Construction Level 2 of the same competition.

Exploration (X-HAB) Innovation Grants in 2015 and 2016 for Mars Transit and Surface Habitats. Working with leading aerospace subject-matter experts and engineers, SEArch+ partnered with architectural design firm Clouds AO in 2015 to win first place in NASA's Phase I Design Competition for a 3D-Printed Habitat for the proposal, Mars Ice House, which was announced at World Maker Faire New York that year.

ICE HABITATS

To date, we have worked on two proposals for Martian ice habitats with NASA. The first is a surface-based design called **Mars Ice House**.

Given the predicted abundance of water in certain areas on Mars, our approach takes advantage of the properties of H_2O as the primary fabrication material. In contrast to Martian habitats that bury astronauts underground in regolith (Martian soil), Ice House introduces water-ice as a superior radiation shield for a long-duration mission, while the translucency of the habitat shell enables the astronaut-crew to establish a connection with the landscape beyond.

It prioritizes a life above ground and celebrates the human presence on the planetary surface.

To construct the habitat, a precision-manufactured ETFE membrane deploys and inflates from a vertically oriented lander storing all mission-specific robotics, materials, and resources. Ice House has been programmed to feature crew quarters for four astronauts, hygiene areas, exercise and medical support, vertical hydroponic gardens, and a wardroom.

In 2016 SEArch+ consulted with NASA Langley Research Center in the design, feasibility study, and risk reduction efforts of our second proposal, **Mars Ice Home**, a related concept for a deployable Mars habitat filled and frozen with indigenous water-ice. Ice Home uses an inflatable membrane filled with water-ice for structural support and radiation protection for a crew. Materials forming the wall assembly of Ice Home have been selected for MISSE-11 (Materials International Space Station Experiment) testing aboard the International Space Station this coming year. Samples will be mounted to the outside of the ISS to investigate the effects of long-term exposure and will then be returned to Earth for analysis.

MARS X-HOUSE

Our team is currently partnering with 3D printing provider Apis Cor for Phase III of NASA's 3D-Printed Habitat Competition, winning first place in Construction Levels 1 and 2, and first place in 100% Virtual Design. **Mars X-House**, our current proposal for the competition, introduces the design and construction sequencing for an autonomously built 3D-printed habitat using indigenous ISRU (in-situ resource utilization) materials to support a crew of four for one Earth-year on a pioneering mission to Mars.

Rather than burying habitats underground or entombing them in regolith, the design of Mars X-House seeks to exceed current radiation standards while safely connecting the crew to natural light and views to the Martian landscape. By vitally connecting the human residents with views to the landscape beyond, the habitat synthesizes key design factors fundamental to future Martian habitation: program and construction efficiency, light, and radiation protection — creating a highly functional and protective habitat for its occupants.

X-House will be 3D-printed at ⅓ scale in the competition's final on-site challenge at the Caterpillar Edwards Research & Demonstration Center. Research in off-world 3D printing is still in its infancy, and it remains indefinite whether the material porosity of 3D-printed regolith may indeed sustain air-tight pressurized enclosures essential for breathable environments supporting human life. The competition's design and construction prototyping levels seek to advance present research in whether a regolith based 3D-printed habitat may indeed enclose a pressure vessel within an off-world environment. The competition's final head-to-head event, planned for May 4, 2019 at Caterpillar's facilities in Peoria, Illinois, asks teams to print the habitat design at scale and conduct an impact test, smoke test, and crush test. The event will be broadcast on NASA TV and visitors are welcome to Caterpillar on the day of the event. ◐

SEArch+, Clouds AO, Apis Cors, NASA Langley

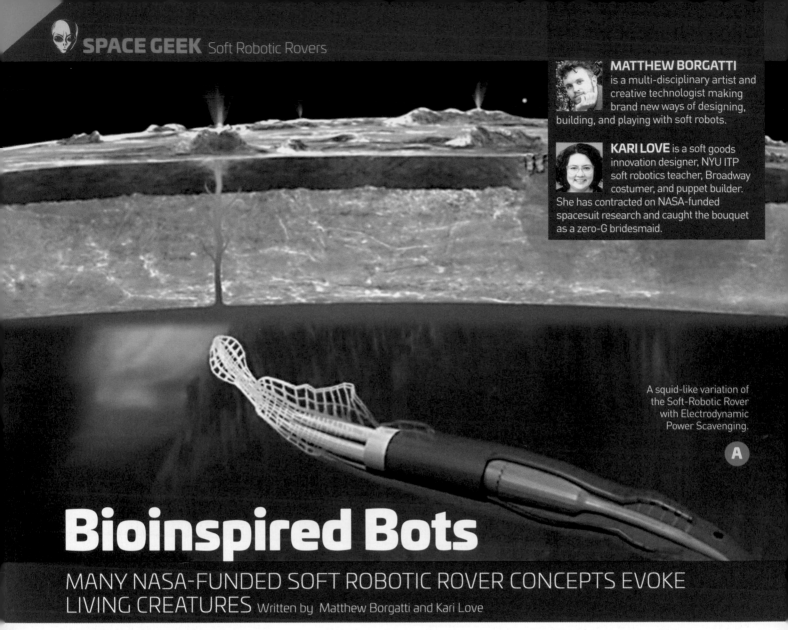

MATTHEW BORGATTI
is a multi-disciplinary artist and creative technologist making brand new ways of designing, building, and playing with soft robots.

KARI LOVE is a soft goods innovation designer, NYU ITP soft robotics teacher, Broadway costumer, and puppet builder. She has contracted on NASA-funded spacesuit research and caught the bouquet as a zero-G bridesmaid.

A squid-like variation of the Soft-Robotic Rover with Electrodynamic Power Scavenging.

A

Bioinspired Bots

MANY NASA-FUNDED SOFT ROBOTIC ROVER CONCEPTS EVOKE LIVING CREATURES Written by Matthew Borgatti and Kari Love

AS NASA-FUNDED SOFT ROBOTIC AND COMPLIANT SYSTEM CONCEPTS BEGIN TO ATTRACT ATTENTION, we jokingly refer to the rover concepts as the "Menagerie" since so many take research inspiration directly from animals. These bioinspired exploration robots include soft elements.

SQUID PROPULSION

Eye-catching with its squid-like form, Mason Peck and Robert Shepherd's Soft-Robotic Rover with Electrodynamic Power Scavenging (Cornell University) serves as a perfect illustration of applying the mechanics of biology to exploration (Figure **A**). This rover can travel where wheeled vehicles aren't suited to the terrain and where solar power and nuclear power are not viable solutions. The moons Europa and Enceladus, whose oceans lie beneath a thick layer of ice, are two potential targets. The concept behind the bioinspired squid

form is to provide efficient motion in a fluid medium using an incredibly low-power system.

This robot's architecture opens up the possibility of escaping dependence on limited-life batteries, and scavenging small amounts of power from locally changing magnetic fields and directly applying it to electrolyzing water. Electrolysis produces a mixture of H_2 and O_2 gas, which can be used to power the robot pneumatically or, even more excitingly, by combustion. When the gas ignites, internal chambers expand causing shape change or even providing direct water-jet propulsion. Although it is still in early research stages, it is not hard to imagine the advantages and appeal of a robotic squid jetting through an alien ocean.

INSECT FEET

If you've ever seen a beetle crawl up a wall or a fly rest effortlessly on the ceiling, you've

seen a complex compliant mechanism in action. Certain insects have lots of tiny hooked hairs called *setae* on their feet (Figure **B**). As a fly brushes its feet across a surface, the hairs flex and hook into microscopic crevices. One on its own could never support the fly's weight, but only a small fraction of them need to catch for the fly to get a firm grip. The hairs are biased in a particular direction so the fly can easily let go by just brushing its foot backward.

When this concept is shifted to the flexible engineering space, the hairs become hinged parts that rely on compliant materials for the flex feature and they have a literal metal hook on the end. At NASA's Jet Propulsion Laboratory (JPL), roboticist Aaron Parness has been experimenting with mass production of such mechanisms.

The hook "hairs" get bundled together into a rover foot, and the result is a rover that can climb upside down and sideways.

These Anchoring Foot Mechanisms have been attached to JPL's LEMUR IIb climbing robots, and tested in conjunction with rock coring drills (Figure). Now, suddenly, missions to other planets and moons can include the option of high-value cliff science. Such systems could also be used for asteroid capture and drilling.

ROVERS THAT BOUNCE, ROLL, OR CLING

The challenges associated with getting a rover onto a planetary surface are huge. Landings typically occur at high speeds and temperatures, relying on single-use hardware and precision timing. New concepts would cleverly circumvent these problems by letting the innate softness of compliant materials absorb the impact. Ideally, the rovers can be launched from orbit, and they will bounce and roll when they contact the planetary surface. The intended impact velocity can be as fast as 15 miles per second without external support, and the rovers will keep on going undamaged. The ability to take a fall without injury also contributes to the possibility of high-value cliff exploration, as was mentioned for the fly-footed climbing robot. This way, rover missions can be carried out on an astronomical body in which a parachute system could never work, such as one with no atmosphere. Other advantages of soft rovers are that they can be packed flat, which maximizes space, and they are relatively light, which drops launch cost.

Tensegrity refers to a structure made of multiple rigid components (none of which touch each other) held together with continuous tension from flexible components. The Super Ball Bot robotic rover harnesses the flexibility of tensegrity structures to traverse challenging terrain (Figure **D**). In this bot, and the other tensegrity projects that have sprung up in its wake, the rigid components shorten and lengthen. The alteration of the center of gravity results in the robot tipping over. The repetition of this tilt generates a bumpy rolling motion, which can be directed toward its target.

These tensegrity rovers have the innate ability to compensate for a single-line component failure of either a hard or flexible part by altering its movement patterns. The flexibility to keep on working beyond component failures leads to a graded capability that can extend missions. Current testing indicates that there can be multiple individual failure points in the tensegrity rover before total loss of useful motion.

How these soft/hard suspensions are bioinspired may be less obvious than other examples from soft robotics. Remember that the bones in our skeletons are held in tension by our ligaments. These aren't the only soft components doing functional work in our skeletal system, either. Cartilage serves as a protective cushion in the space between our bones, and tendons attach muscle to bone.

Another concept for tensegrity planetary exploration are the Moballs from the California Institute of Technology. These bouncing balls would pack flat for launch and would then inflate and rigidize when deployed. On the planet's surface, they would be semipassive, collecting kinetic energy from wind or downhill slopes to discharge as magnetic force to add stopping ability and directional driving. Their form takes inspiration from passive biological systems such as tumbleweeds.

Both genres of rolling rover concepts aim to take the network advantage of a small and distributed technology, similar to what we see from CubeSat miniaturized satellites, and bring it to the ground level. By making them relatively small and cheap, you can send more robots, cover more ground, and take more risks, collecting a larger array of data. More data means broader scientific opportunities.

Using an entirely different mode of locomotion, the AoES (Area-of-Effect Soft-Bots) by Jay McMahon of the University of Colorado, Boulder, is specifically designed for the environment on and around rubble-pile asteroids. This bot takes advantage of the large, flexible surface of this type of asteroid to gain adhesion-based anchoring, surface mobility via crawling without pushing itself off the asteroid, and a fuel-free orbit and hopping control using solar radiation pressure (SRP) forces. By adding the options of clinging to the surface, hopping, or orbiting, these soft robots aim to perform on low-gravity asteroids that can be unstable due to their low density. ●

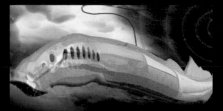

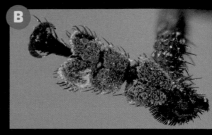

Proposed undersea rover for exploring moons Europa (Jupiter) and Enceladus (Saturn).

B

Close-up of a weevil foot showing the setae, which help the insect cling to surfaces.

C

LEMUR IIb climbing robot using a bioinspired gripper to hang from a model Martian rock.

D

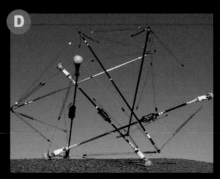

The Super Ball Bot is an all-in-one landing and mobility platform based on tensegrity structures, which allows for lower-cost and more reliable planetary missions.

NASA/Cornell University, NASA/National Science Foundation, NASA/JPL-Caltech, NASA

Cosmic Cosplay

Written by Sophy Wong

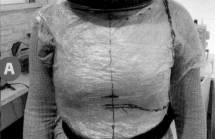

TIME REQUIRED:
A Weekend

DIFFICULTY:
Intermediate

COST:
$30–$50 + Laser Cutting

MATERIALS
- » EVA foam, 6mm thick
- » Plastic wrap
- » Blue painter's tape
- » Contact cement
- » Costume or helmet to add your embellishments to
- » ModPodge or wood glue (optional)
- » Primer and paints (optional) You can use tube acrylics, spray paints, or an airbrush.

TOOLS
- » Laser cutter
- » Flatbed scanner
- » Computer with vector graphics software such as Adobe Illustrator, Inkscape, etc.
- » Camera
- » Scissors or craft knife
- » Heat gun or hair dryer
- » Sanding pad, ultra fine
- » Marker pen

SOPHY WONG
is a designer and maker whose projects range from period costumes to Arduino-driven wearable tech. She can be found at sophywong.com and on her YouTube channel chronicling her adventures in making.

A

B

CUSTOMIZE ANY COSTUME WITH REALISTIC DETAILS MADE FROM LASER-CUT EVA FOAM

IT'S NO SECRET: COSPLAYERS ARE USING EVA FOAM TO CREATE INCREDIBLE COSTUMES AND PROPS, and sharing their techniques so that everyone can learn how to use this versatile material. A little paint and a lot of patience can turn EVA foam into armor, a magical sword, or even a spacesuit! Let's look at how EVA foam can transform a jumpsuit and a purchased helmet into a spacesuit fit for a sci-fi thriller.

With practice, concentration, and a very sharp blade, it's possible to achieve precise results when cutting EVA foam by hand. Another option is to cut your foam in a laser cutter, which is how the costume shown here was made. The laser cutter can score lines, engrave areas, and make through-cuts that are extremely clean and precise — perfect for sci-fi spacesuit details!

Cutting EVA foam in a laser cutter is a bit of an experimental technique, and you will need to do some tests to determine the optimal power and speed combination for your foam. The laser may produce a wider kerf on foam than on other materials, as the cut edges shrink away from the heat of the beam.

> **CAUTION:** Always double-check the material's packaging label to make sure that your foam is indeed EVA (ethylene-vinyl acetate) before cutting. Some other vinyl materials such as PVC (polyvinyl chloride) will release chlorine gas when cut and are not safe to put in a laser cutter. Check with the manufacturer of your laser cutter for a list of safe materials.

1. MAKE YOUR PATTERN

Start with the tried-and-true cosplay method of plastic wrap and blue painter's tape. For large areas like the torso, Press'n Seal plastic wrap is a great alternative to using tape. While wearing your jumpsuit, or similar clothing, cover the area of your body you want to pattern with a layer of plastic wrap, then cover the plastic wrap with pieces of blue painter's tape or Press'n Seal in a solid layer (Figure A). For a symmetrical design, you only need to do one half of the costume. You can use the same process for rigid costume pieces like a helmet (Figure B).

Now use a marker to draw the shapes of your foam garments or embellishments directly onto the blue tape or Press'n Seal (Figure C). Keep in mind that your foam will start out flat, and can be somewhat heat formed into a curve after cutting. To wrap foam completely around the body, as in the torso, you will need to add darts.

Once your design is drawn out, label each piece so you know what to cut and where it goes. Take a photo of your design for reference later, and peel off the plastic wrap and tape together in one layer (Figure D). Cut out each pattern piece with scissors or a craft knife. Mark and cut darts into your pattern pieces as needed to make them lie flat (Figure E). Once cut, you will glue the darts closed to achieve smooth curves.

2. DIGITIZE YOUR PATTERN

Once flat, use a flatbed scanner to scan the pattern pieces into your computer. Scan them as images at 300 dots per inch (dpi) (Figure F).

Open each scanned image in a vector program like Adobe Illustrator and trace the shapes to create vector outlines (Figure G). The Bézier curve/pen tool works great for this. Remember that straight lines will curve and distort over a spherical surface.

When your shapes are outlined, add any decoration inside the shapes you like. The costume shown here uses several different textures to create the illusion of different materials, even though they're all made of the same foam. Try adding score lines and engraved patterns to some pieces, while leaving others plain (Figure H). Have fun with this, and make your spacesuit truly unique!

Once your vector shapes are designed, do a check by printing them out on regular paper at actual size, using the same resolution you scanned them at (300 dpi). Cut the pieces out and tape them in place on your costume to make sure you like the sizes and shapes you've made (Figure I on the following page). Make any adjustments at this stage before cutting your final foam material. To account for the thickness of the foam and shrinking during

Kim Pimmel, Sophy Wong

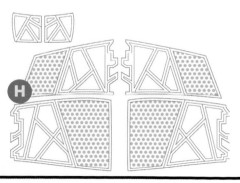

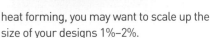

Kim Pimmel, Sophy Wong

heat forming, you may want to scale up the size of your designs 1%–2%.

3. CUT THE EVA FOAM

Refer to the instructions for your laser cutter for how to set up your file for cutting, scoring, and engraving, as methods vary by machine. Work out your power and speed settings by cutting test pieces first. Ideally, use the highest effective speed and the lowest effective power. Foam can melt and ball up with intense heat, so when cutting aim for two passes at low power rather than a single pass at high power (Figure **J**).

The foam used in this project was 6mm EVA foam from TNT Cosplay Supply. It comes on a roll, and should be flattened out before cutting. To heat seal and flatten the foam, run a heat gun over both sides of the foam until shiny.

Here are the settings that were used to cut, score, and engrave the pieces of the costume shown (Figures **K** and **L**):
- **Laser cutter:** Glowforge Basic
- **Material:** 6mm EVA foam
- **Cut:** Speed: 300 / Power: 80 / Passes: 2
- **Score:** Speed: 500 / Power: 20 / Passes: 1
- **Engrave:** Speed: 1000 / Power: 60 / Passes: 1 / LPI: 125

4. ATTACH THE PIECES

Before attaching the EVA pieces to any smooth plastic surfaces, roughen the plastic slightly with an ultra fine sanding pad. This will give your glue and paint more purchase. (Remove the visor of your helmet and set it aside to keep it scratch free.)

Glue any darts/joins closed and heat form as needed to create the curves of your costume. When applying additional layers of foam details, hold each foam piece in place on the costume (Figure **M**) and trace around it with a pencil to create outlines so you know exactly where to put the glue. If any foam pieces need to be curved, apply a little heat with a heat gun and work them into shape with your hands. When working with contact cement, apply cement to both surfaces to be joined and let dry before laying each piece in place. Once the glued surfaces touch, they will be permanently joined (Figure **N**).

5. SEAL AND PAINT

Before painting, seal the foam with a few quick blasts from your heat gun — just enough to turn the surface of the foam smooth and a little shiny. This will also widen and enhance any scored lines and engraved patterns. Be careful not to melt or distort the plastic helmet.

For an extra smooth surface, apply several layers of ModPodge or wood glue and let dry before sanding, priming, and painting with tube acrylics, spray paint, or an airbrush (Figures **O** and **P**).

GOING FURTHER

For truly expert guidance on painting, weathering, and foam construction, check out Punished Props Academy, Kamui Cosplay, and Evil Ted Smith. They're just a few of the talented makers teaching foam crafting techniques via YouTube. After watching their inspiring and informative videos, you'll be ready to tackle your own foam masterpiece! ●

Galactic Garments

Written by Caleb Kraft

FIND COSPLAY INSPIRATION IN THESE SPACE-THEMED FASHIONS

You may not be in line for the next stint on the International Space Station, but you can still dream for — and dress for — the job you want. People have been creating costumes around space travel and space fantasy themes since long before we could actually visit the cold expanses. Here are a few variations for your inspiration.

BUDGET BLAST-OFF

Let's say you've got a craving for some astronaut time, but you don't have the cash to go all out on your suit. Audrey Love's budget-friendly version uses papier-mâché for the helmet and cheap paper-based coveralls for the body. You'll spend some time constructing the helmet but your wallet won't be too much lighter. For a dirt-cheap build, the results are actually pretty good looking from across the room, especially once you add the LED mood lighting. www.instructables. com/id/Astronaut-Costume/

WELL-SUITED

Ryan Nagata is a cosplayer with an eye for detail. So much so that celebrities such as Adam Savage have hired him to build suits. He's even been brought in on some major films you've probably seen, such as *First Man*. His personal obsession with building the ultimate spacesuit has taken over, and now this is what he does full time. It would be safe to say that he's gone pro.

Specializing in faithful reproductions of past NASA suits, he painstakingly researches materials and methods to ensure that his builds are perfect down to the last fiber. ryannagata.com/ portfolio/spacesuits

NO ONE CAN HEAR YOU SCREAM

Of course, space cosplay doesn't always have to include a spacesuit and helmet. Look at this fantastic example of a reproduction from the movie *Alien*, made by Andy Poulastides. This is the jacket the crew of the ill-fated *Nostromo* wears. Re-creating this will have you flexing your needle and thread skills, and you'll have to source a few custom patches, but the results will win you some major geek cred at any event. therpf. com/forums/threads/my-nostromo-crew-jackets-pic-heavy.293168

Audrey Love, Ryan Nagata, Andy Poulastides

Edge-Lit **LED Rainbow**

Explore light piping and color mixing in edge-lit acrylic with this fun IoT weather display Written and photographed by Debra Ansell

DEBRA ANSELL
(geekmomprojects.com) believes that LEDs improve everything, particularly wearable tech. She has a passion for making STEM education accessible to all.

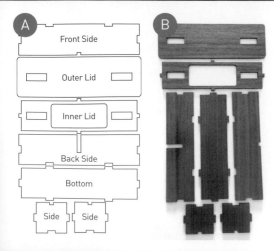

Light prefers to take the shortest path, but under the right conditions you can make it bend and bounce, generating lovely effects in the process. This project takes advantage of some special properties of acrylic. Inside polished acrylic, light rays reflect off the smooth inner surfaces, preferring to escape at the rougher edges or scatter at imperfections in the surface. This internal reflection causes the light to suffuse the entire acrylic shape with a subtle uniform glow, even where it bends and curves.

We can take advantage of light scattering by engraving patterns into the smooth surface of acrylic sheet then illuminating the sheet to generate a brighter glow at the etched spots. These edge-lighting effects are often seen in signage or décor used in dark venues (see "LED Nixie," Make: *Volume 66, makezine.*

com/projects/led-nixie-display).

The edge-lit rainbow juxtaposes seven separate acrylic "arcs" and illuminates each one individually with RGB LEDs so that each arc glows with its own colors, independent of its neighbors. This project takes the effect one step further, giving both ends of each arc separate RGB LEDs to generate a color mixing effect.

There are many possible ways to control RGB LEDs, so I made this project versatile by attaching the LED strips to a JST connector, allowing it to be easily powered by interchangeable microcontrollers. This build explores ways to illuminate the rainbow using CircuitPython code on a Circuit Playground Express for quick programming and interactive color-mixing, as well as an ESP8266 board for IoT connectivity and a colorful report on real-time weather conditions.

1. DOWNLOAD THE VECTOR FILES

Download the free vector files for this project at makezine.com/go/edge-lit-rainbow in SVG or DXF format. If you have access to a laser cutter, you can cut them yourself; otherwise an online service like Ponoko can cut them for you. The outlines are sized precisely to fit the LED strips, so it's important not to change the scale of the vector files before cutting the materials.

2. LASER-CUT AND ASSEMBLE THE ENCLOSURE

Use the design in *WoodBox.svg* (Figure Ⓐ) to cut the pieces of the box enclosure from ⅛" wood (Figure Ⓑ). Delete the red text labels from the file before cutting.

Use wood glue to attach the inner and outer lid pieces together (Figure Ⓒ) so that the small rectangular holes in both pieces align as closely as possible. While the glue is drying, clamp or tape the pieces together with painter's tape to maintain their alignment.

Lightly glue the interior edges of the enclosure body, and assemble the pieces using the slots and tabs for alignment and stability (Figure Ⓓ). While the glue dries, clamp or tape the pieces of the enclosure body in place.

3. LASER-CUT THE ACRYLIC

There are three different vector files for the acrylic pieces. Cut the file *Rainbow.svg* (Figure Ⓔ) from ¼" clear acrylic. The red lines in the file are for etching, and black are for cutting. Some of the acrylic pieces have narrow parts that can be brittle if stressed, so be sure your laser power settings make a clean cut all the way through. The small notches in the ends of each arc may stick a little, but can be easily dislodged with just a tiny amount of pressure from a fingernail.

Cut the clouds from ⅛" translucent blue acrylic using *Clouds.svg* (Figure Ⓕ) and the two support pieces in *Support.svg* (Figure Ⓖ) from ⅛" clear acrylic. Peel off any protective masking paper, leaving just the bare acrylic (Figure Ⓗ).

4. APPLY COPPER TAPE

Placing strips of copper tape between the rainbow arcs will reflect any light that tries to escape from the interior acrylic edges. The tape also helps to separate the different LED colors in the

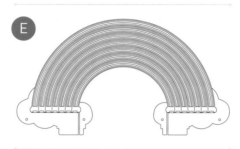

TIME REQUIRED:
2–3 Hours

DIFFICULTY:
Intermediate

COST:
$50–$100

MATERIALS

» **Circuit Playground Express or Adafruit Feather Huzzah ESP8266** or other microcontroller capable of controlling RGB LED strips. For the full weather display functionality, you'll want the Huzzah or another microcontroller with Wi-Fi.

» **RGB LED strip, SK6812 type, 3535 size, white** with a pixel density of 144 LEDs/meter, such as Adafruit #2969 (adafruit.com) or SparkFun #14732 (sparkfun.com)

» **Clear acrylic sheet, ¼" thick, 6"×10" or larger**

» **Clear acrylic sheet, ⅛" thick, 5"×5" or larger**

» **Translucent blue acrylic sheet, ⅛" thick, 8"×5" or larger**

» **Wood sheet, ⅛" thick, 8"×12" or larger** for the laser-cut enclosure. I found a nice selection of laserable wood sheets at Johnson Plastics johnsonplastics.com/engraving/sheet/laser/wood and used a lovely walnut veneer for this project.

» **Copper tape, ³⁄₁₆"**

» **Wood glue**

» **Machine screws, M2×16mm (4)** with nuts

» **Hookup wire, 26 AWG solid insulated, white**

» **Heat-shrink tubing**

OPTIONAL:

» **JST connectors, 4-pin: male (1) and female (1)**

» **Painter's tape** very helpful

» **Crimp ring terminals, M3 uninsulated (4)**

» **JB Weld epoxy** or similar glue that can adhere to plastic and wood

» **Sharpie markers: red, black, and blue**

TOOLS

» **Soldering iron and solder**

» **Laser cutter (optional)** You can also outsource the laser cutting.

» **Wire cutters/strippers/pliers**

» **Computer with Arduino IDE or Mu code editor**

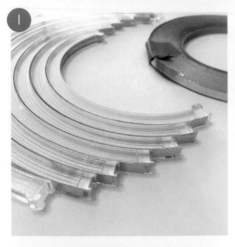

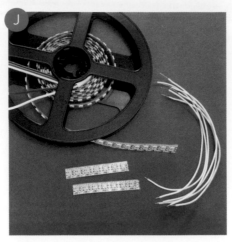

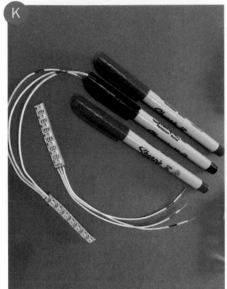

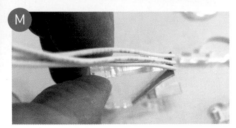

acrylic arcs, minimizing color bleed and keeping the colors sharp and distinct.

Cover the *top* rounded edge of the smallest six arcs (but not the outermost arc) with a piece of ³⁄₁₆" copper tape (Figure **I**). Don't let the tape extend past the end of each arc, or you'll risk a short-circuit if the copper tape comes in contact with the LED strips.

5. PREPARE THE LED STRIPS

Cut two segments containing 7 LEDs apiece from the SK6812 3535 RGB LED strip. Cut six 5" lengths of white 26 AWG hookup wire and strip a bit of the insulation from the ends (Figure **J**). Using white insulation makes the wires less obvious through the transparent acrylic. It also makes it harder to distinguish between the leads for 5V, signal, and ground, so we'll mark them once they're connected.

Tin one end of each piece of wire, and solder connections to the three input pads of each cut LED strip. Mark the end of each wire with a colored Sharpie, denoting its purpose: 5V red, Signal blue, and GND black (Figure **K**). Using pliers, carefully bend the wires down from the LEDs just past the solder joints. The length of LED strip should fit, with a little extra space, along the long straight edge of the ¼" acrylic cloud piece, as shown in Figure **L**. The wires should bend just at the corner of the acrylic piece.

6. ATTACH THE LED STRIPS

Remove the adhesive backing from the LED strips, and stick them firmly to the long straight edge of the ¼"-thick clear cloud piece. The unsoldered edge of the LED strip should extend along the edge to just about the place where the curved notch starts. It's likely that the LED strip is wider than ¼", so align the back side of the strip (the side furthest from the viewer) to one side of the straight edge and allow the front edge of the LED strip to overhang the other side of the acrylic a bit (Figure **M**).

7. STACK THE CLOUD PIECES

On each side of the rainbow, set out the cloud pieces as shown in Figure **N**. Each cloud consists of four layers. The bottommost layer is the blue cloud shown nearest the rainbow, and each acrylic piece will stack over the one shown just above it in the picture.

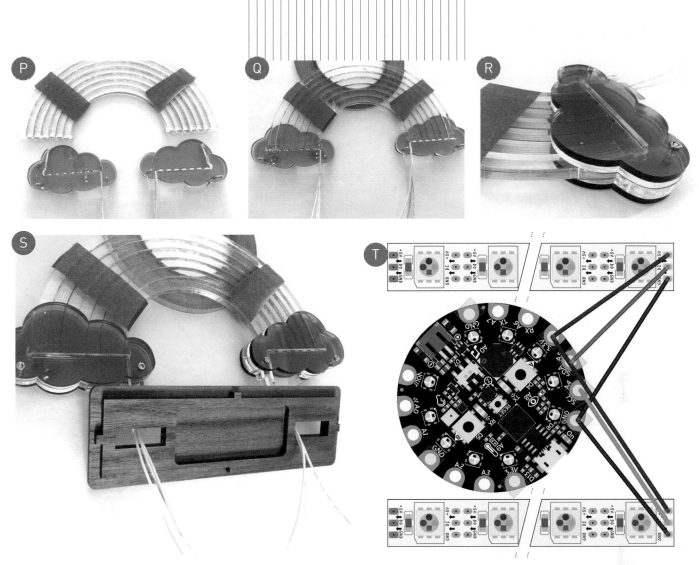

Take the 16mm M2 screws and insert them through the holes in the first blue cloud as shown. Place the cloud on a level surface so that the screws point upward.

Next, stack the two ¼" acrylic cloud pieces over the blue cloud so that the M2 screws pass through the hole in each piece, and thread the hookup wires down through the narrow slot between the two ¼" pieces (Figure O).

The rainbow also sits in this layer. Place the seven rainbow arc pieces together on a flat surface with the notched edges forming a line. Be sure each arc is resting with its etched face oriented downward. Apply a few small strips of painter's or masking tape to help hold all seven arcs in alignment (Figure P). Lower the ends of the rainbow into the wide gaps between the two ¼" pieces of acrylic cloud (Figure Q). The small bumps on the innermost and outermost arcs should just fit into the corresponding curved notches in the ¼" cloud pieces. If the pieces don't fit easily, don't force them. Instead, try *very slightly*

rotating the ¼" acrylic cloud pieces until the fit is good.

Once in place, support the center of the rainbow from underneath so the ends won't dislodge. (The ³⁄₁₆" roll of copper tape works nicely for the task, if you place it under the tops of the arcs.) The rainbow arcs should be seated inside the clouds so that each LED is nestled in the gap at the bottom of an acrylic arc. If the LEDs and arcs don't align, you may have to remove the rainbow arcs and re-seat the LED strip.

Once the rainbow arcs are in place, cover the bottom edges with the ⅛" clear support pieces by sliding them over the M2 screws. These should sit so that the horizontal slot just encompasses the overhanging edge of the LED strip from Step 6. This allows the acrylic cloud to be sandwiched snugly together without squishing the LED strip.

Finally, add the last blue acrylic cloud to the top of the stack, and screw down the M2 nuts to hold the pieces together (Figure R). The nuts currently sit on the front side of the clouds. Later, we can reverse the

screws, but for now, they'll hold the acrylic pieces nicely in place.

8. ATTACH RAINBOW TO ENCLOSURE
Take the top of the wooden enclosure and run the white hookup wires through the corresponding rectangular holes (Figure S). Slot the rectangular bottom portions of the clouds through the same holes. They will fit loosely. Once the electronics are successfully tested, you can glue the rectangular acrylic tabs into the slots for greater stability using JB Weld or similar adhesive.

9. SOLDER THE CONNECTOR
I like to solder a detachable JST connector to the ends of my LED strips, so that I have the flexibility to swap microcontrollers without resoldering the connections. If you do choose to solder the leads directly to a controller, make the connections shown in Figure T. The instructions below describe how to attach a 4-pin JST cable connector to your wires.

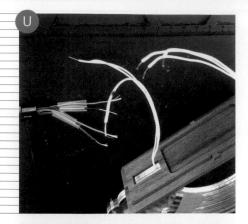

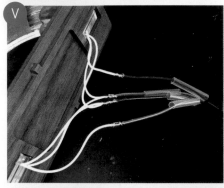

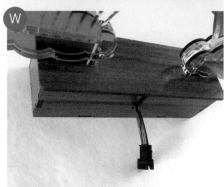

Take the color-marked hookup wires and bend them so they meet near the middle of the enclosure lid. Trim the wires from the male JST connector to a few inches in length and strip the ends. Slide heat-shrink tubing over each wire (Figure U).

Each LED strip signal wire has its own wire in the JST connector. The 5V wires both connect to the red JST connector wire, and the two ground wires both connect to the black JST connector wire. Trim the white hookup wires so they meet in the middle and extend an inch or so away from the lid, then strip the ends. Be careful not to lose track of which wire is which — you may want to mark the color of each wire again with a Sharpie.

Twist the ends of the hookup wires together with their corresponding connector wire as seen in Figure V. Solder the connections, then slide the shrink tube over each junction. Heat the shrink tubing so that it shrinks to reinforce and insulate each join.

Now place the lid on top of the enclosure base, with the JST connector extending out the slot in the back of the enclosure (Figure W). The tabs in the underside of the lid fit in the slots at the top edges of the enclosure.

10. CONNECT THE CIRCUIT PLAYGROUND EXPRESS

An Adafruit Circuit Playground Express (CPX) running CircuitPython will get your LED rainbow up and running quickly. Its built-in buttons and sensors make it easy to add interactive features to the program.

Connect a female 4-pin JST connector to the CPX. You can solder the connections shown in Figure T directly, or, if you have some M3 crimp ring terminals, you can strip the ends of the JST connector wires, crimp the ring terminals over the ends, then use M3 screws and nuts to secure the connector to the circular pins on the CPX (Figure X).

11. MAKE IT GLOW WITH CIRCUITPYTHON

The CircuitPython code (file *main.py*) uses the CPX's built-in switch, buttons, and accelerometer to change color effects in response to user interactions. The code uses the *FancyLED* libraries by Phillip Burgess (learn.adafruit.com/fancyled-library-for-circuitpython) which accept HSV (Hue, Saturation, Value) colors and make cycling through different hues as simple as counting (Figure Y).

The program runs in two modes, Spectrum and Color Mixing, which are determined by the switch position. In both modes, the colors of each rainbow arc change gradually with time, following a base hue that cycles through the color spectrum.

Pressing CPX button A changes the rate at which the base hue evolves. The rate cycles through four different values — ranging from barely perceptible change to zippy color rotation. Pressing CPX button B toggles the colors displayed at either end of each arc. In Spectrum mode, the hues at either end of each arc are either the same, lighting the arc with a single, distinct color (Figure Z) or they are opposites on the color wheel, generating a color gradient along each arc (Figure Aa).

In Color Mixing mode, the rainbow displays two opposite hues in each arc. These hues evolve with time, and the rate at which they cycle is set by pressing CPX button A. The relative intensity of these two colors responds to the x-axis orientation of the CPX. Tilting the CPX to the left or right makes each color brighter or dimmer (Figures Bb and Cc). Pressing button B in Color Mixing mode toggles the colors on each side from uniform, to interlaced (Figures Dd and Ee).

12. MAKE IT A WEATHER DISPLAY WITH AN ESP8266

A Wi-Fi-enabled board, like the Adafruit Feather Huzzah ESP8266, can turn your edge-lit rainbow into an IoT weather display. Different colors in the rainbow arcs represent the temperature, and lively animations indicate weather conditions like rain, snow, or strong winds.

If you terminated your rainbow LED wires with a male 4-pin JST connector, then connect the Feather to the corresponding female JST connector. Make the connections shown in Figure Ff. I soldered female jumper wires to the four female JST connector wires, and attached them to header pins on the Feather ESP8266 (Figure Gg).

Download the Arduino code from the project page at makezine.com/go/edge-lit-rainbow. It's based on Adafruit's ESP8266 Wi-Fi Weather Station Learning Guide (learn.adafruit.com/wifi-weather-

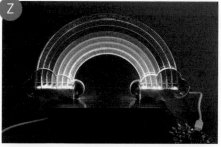

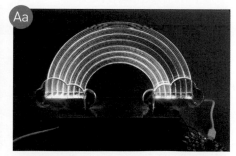

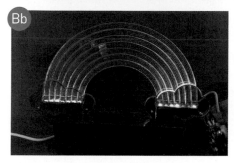

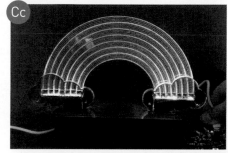

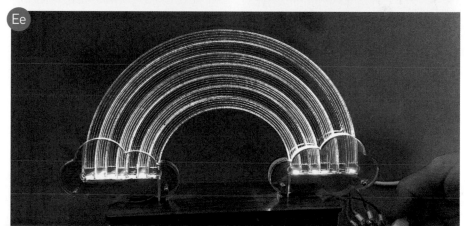

station-with-tft-display) which uses OpenWeatherMap — a free online service with an API that allows you to check local weather conditions as often as every 10 minutes.

You'll need to create an account with OpenWeatherMap and get an API key at docs.thingpulse.com/how-tos/openweathermap-key. Once that's done, find the numerical code for your location by going to openweathermap.org/find, searching for your city and clicking the link to see its current weather. The numerical city code will be the last part of the URL.

Replace the placeholder values in the Arduino sketch with your API key and city code, along with your Wi-Fi SSID and password. Compile and upload the code. The LEDs will display an evolving rainbow spectrum, which, every 30 seconds, will glow uniformly in a solid color to represent the current temperature. Additional animations indicate weather conditions, like arc colors dripping downward for rain, sparkling for snow, and sliding side to side to indicate strong winds.

Place your rainbow somewhere you'll be able to see it when you want to know the weather. Even ugly conditions will look attractive now! ⊘

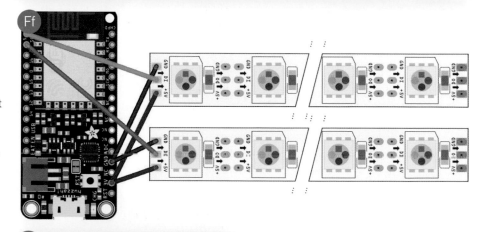

Use blue "sun print" paper
to take unique UV photos
with a cardboard camera

Written and photographed by Josh Lewis

DIY
Cyanotype
Camera

JOSH LEWIS is an
amateur photographer,
professional geek, and
trained artist. He is also
a father, a cook, and
sometimes a teacher.

House, imaged with Fresnel lens.

Deer skull, imaged with Fresnel lens.

House, imaged with magnifying glass.

Deer skull, imaged with magnifying glass.

Here's how to make a simple sliding box camera that takes long-exposure photos on cyanotype paper that's normally used for blueprints or "sun prints." The paper can be easily developed at home in a bathroom sink. It's a fun way to really, physically see what's happening in a camera. It's an inexpensive project, and it produces wonderfully spooky UV images.

Before you build your camera, play with the cyanotype paper. Do some experiments with the sun print kit and see how the paper works in direct sunlight. Explore UV exposure by writing with sunscreen on the included acrylic sheet and placing that over the paper. Become familiar with developing the paper, and try using drugstore hydrogen peroxide to speed the oxidation process after development.

1. CHOOSE A LENS
You want a large, inexpensive, thin lens for this project: maybe a Fresnel lens from Amazon or eBay, or a large drugstore magnifying glass. The Fresnel lens (shown on opposite page) is large and gathers lots of light, making exposures not so long (30 to 90 minutes), as shown in Figures A and B. A magnifying glass (shown on this page) will help form a clearer picture (Figures C and D) but might require longer exposure times (1 to 3 hours).

2. MEASURE ITS FOCAL LENGTH
In a darkened room with a window, use your lens to focus an image on the wall — an image of something far outside the window. Use a ruler or yardstick to find out how far away the lens is from the wall; that is approximately the focal distance. For a demonstration, watch youtu.be/joQw3_rM_jl.

3. BUILD BOX SLIDE AND ATTACH LENS
My lens had a focal length of 300mm (11.8"). I built a sliding attachment that was about 10"×10"×11" and attached the lens to its front. This sliding portion slides in and out of the box so you can focus the image. I also allowed the top of this sliding portion to fold open (Figure E) so I could peek into the box and see the image while focusing.

4. RE-MEASURE FOCAL LENGTH
When you have the lens attached to the box slide, measure the focal length again with everything squared up (Figure F). Can you

TIME REQUIRED:
A Weekend

DIFFICULTY:
Intermediate (OK for Beginners)

COST:
$20–$60

MATERIALS
» **Lens** either a large magnifying glass lens or a Fresnel lens
» **Cardboard boxes**
» **Cyanotype paper** I use the Super Sunprint Kit from Lawrence Hall of Science, Amazon #B001KOGY3M or American Science & Surplus #93519P1.
» **Tape** I recommend gaffer's tape.
OPTIONAL:
» **White glue** such as Elmer's
» **Paper**
» **Hydrogen peroxide, 3%**

TOOLS
» **Box cutter**
» **Ruler**
OPTIONAL:
» **Paintbrush and bowl, or spray bottle**
» **Scanner** if you want to create sienna- or sepia-tone positive images

Camera is secure on the ground. Direct sunlight is not hitting lens. Lens cap is used to shield the opening in the top of the camera.

see the far images clearly? Do you have a little clearance? Cut back the box slide as needed. The box can and should be a little shorter than the lenses' focal length.

5. BUILD THE OUTER BOX

Build a box around the box slide. I found that a 10"×10"×11" box was about right to expose half sheets of the cyanotype paper. I made my slide to fit into it, but you could do the reverse and just build a box around the slide. I tore back the very top of the box to help peek inside the camera while I'm focusing.

6. MAKE A LENS CAP

I took a size 1AD Amazon box that fit over the end of my camera and used it as a lens cap. It protects the lens and prevents accidental exposure. Putting it on when I'm done taking the picture stops exposure.

7. BUILD A WEDGE (OPTIONAL)

Build a little cardboard wedge to go under the camera. The camera is most stable on the ground; the wedge can prop it up to take pictures of buildings and other tall objects.

8. TAKE A PICTURE

Go outdoors and look for a subject. The cyanotype paper responds to UV-A light so it might not produce an image exactly like what you see visually. Look for high-contrast, simple compositions.

> **CAUTION:** Make sure the camera is pointed away from direct sunlight. You have made a cardboard box with a lens on the front — this can be a fire hazard. Always point the lens away from the sun.

Set up your camera (Figure G) and focus your image. Start by making some test exposures: Leave the camera alone ½ hour for a Fresnel lens, 1 hour for a magnifying glass. When the paper is exposed, you can actually see a positive image in the back of the camera, as if the UV light bleached the paper (Figure H).

9. DEVELOP THE PICTURE

In a room without direct sunlight, soak and develop the cyanotype paper in a sink. Rinsing the paper 2 or 3 times helps to remove unreacted chemicals and improve the contrast and lifespan of the print.

A final wash or spray with hydrogen peroxide (Figure I) helps to speed

GIVE A GIFT.
ONE YEAR ONLY $39.99.

Make:

GIFT FROM

NAME _____ (please print)

ADDRESS/APT. _____

CITY/STATE/ ZIP _____

COUNTRY _____

EMAIL ADDRESS (required for order confirmation) _____

☐ Please send me my own subscription of Make: 1 year for $39.99.

GIFT TO

NAME _____ (please print)

ADDRESS/APT. _____

CITY/STATE/ ZIP _____

COUNTRY _____

EMAIL ADDRESS _____

493GS1

We'll send a card announcing your gift. Make: currently publishes 6 issues annually. Occasional double issues may count as 2 of the annual 6 issues. Allow 4-6 weeks for delivery of your first issue. For Canada, add $9 US funds only. For orders outside the US and Canada, add $15 US funds only.

oxidation, darken the blues, and show the full picture you managed to capture. If you don't have hydrogen peroxide you might have to wait up to a day to see if you have captured a strong image.

FROM ULTRAVIOLET TO PRUSSIAN BLUE

Enjoy this simple camera! I've had good luck getting spooky pictures with it. Because they're exposed by ultraviolet light, they have a ghostly look. It's not quite what you're used to seeing.

How does it work? The sun print paper is treated with chemicals that, when exposed to UV light, react to form the pigment Prussian blue (ferric ferrocyanide), also known as Berlin blue or Paris blue. Prussian blue is insoluble in water, so it stays behind while the unreacted chemicals are rinsed away.

INVERT TO SEPIA

If you have a scanner, you can scan your cyanotype picture and invert it digitally using a photo editing program to make sienna- or sepia-toned positive images such as Figures J, K, and L.

If you don't have a scanner, you might try a smartphone and a free photo-editing app like Snapseed. Take a picture of your cyanotype, then edit with Snapseed. Do something like Tools→Curves and change the diagonal curve line so it goes to opposite corners (Figure M). Then try other effects to get a better sepia or black-and-white image.

SEE LIKE A BEE

If you get really good at this you might use a camera extension and very long exposures to see if you can image the UV patterns on flowers. See kennethleegallery.com/html/tech/bellows.php to consider how much extension and how long an exposure you might need. ◐

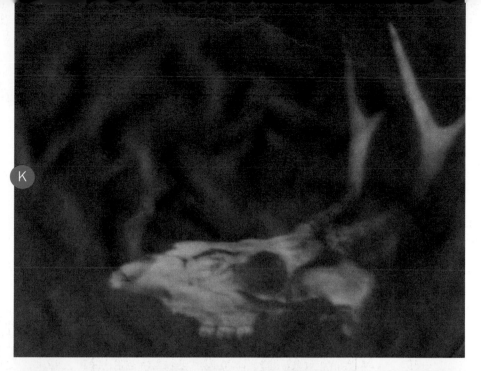

K

L

M

[+] Learn More:
About cyanotypes:
wikipedia.org/wiki/Cyanotype
Cyanotypes for teaching about UV exposure:
makezine.com/go/cyanotype-UV-exposure

Cool beans

TIME REQUIRED:
Half a Day

DIFFICULTY:
Easy

COST:
$25–$35

MATERIALS

» **Wall transformer, 12V 1.5A**
» **Bolts, ¼"×2½", fully threaded (4)** with nuts
» **Washers, ¼" ID (12)**
» **Cooling fan, 12V, 50mm square**
» **Aluminum finned heat sink, approximately 70mm×70mm×25mm** You likely won't find one exactly this size, but anything close will work.
» **Aluminum strips, 1¼"×½"×0.019" thick (4)** You can buy aluminum sheet metal at the hardware store and cut it to size with tin snips. Round and file all edges for safety.
» **Switch, double-pole double-throw (DPDT)** aka "on-off-on" switch
» **Project box, about 1½"×2"×2½"**
» **Hookup wire, 22 gauge insulated, red and black (2' of each)**
» **Peltier thermoelectric modules, 40mm×40mm, 12V 6A (2)** Peltier modules utilize the Peltier effect to heat and cool. They're made from two ceramic plates placed on opposite sides of an array of semiconductors.
» **Thermal adhesive** a small tube

TOOLS

» **Drill with 5/16" and ½" bits**
» **Wire cutter/stripper**
» **Small adjustable wrenches (2) and/or screwdriver** to match your bolt heads
» **Tin snips**
» **File**

Jean Charles Athanase Peltier
and the
Peltier Effect

Build a coaster to keep your coffee warm — or cold! — by flipping a switch **Written by William Gurstelle**

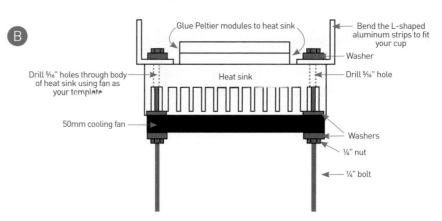

A

Metal type 2

Cold junction

Hot junction

Metal type 1

Metal type 1

Current

Voltage

B

Glue Peltier modules to heat sink

Bend the L-shaped aluminum strips to fit your cup

Washer

Drill ⁵⁄₁₆" holes through body of heat sink using fan as your template

Heat sink

Drill ⁵⁄₁₆" hole

50mm cooling fan

Washers

¼" nut

¼" bolt

In 1798, Jean Charles Athanase Peltier was just 13 years old and although he came from a poorly educated family in rural France, people already were beginning to notice the youngster's intellectual talents.

Besides being an avid reader of nearly any book he could get hold of, Peltier showed an aptitude for fixing clocks. His family was too poor to continue his general education so his father apprenticed him to a clockmaker. Young Peltier found his master, Monsieur Brown, to be extremely unpleasant and extraordinarily controlling. Brown forbade Peltier to educate himself in anything apart from clockmaking. But at night Peltier would clandestinely read by candlelight. That is, until Brown discovered him and removed all candles from his room. Even then, Peltier attempted to read at his window by the light of the full moon. Again Brown found him out and stopped him from doing even that. That was the last straw: "Ça suffit!" Peltier cried and stole away for Paris.

Life was better in Paris. He found work with perhaps the best-regarded watchmaker in all of France: Abraham Louis Brequet. Peltier worked steadily for Brequet until 1815 when he came into a sizeable inheritance through his wife's mother. This financial freedom gave him the opportunity to move beyond clockmaking. For the first time in his life, Peltier could devote himself to what he loved most: learning. He read everything he could get his hands upon: the fiction of Voltaire, the philosophy of Rousseau, but most of all, books on science.

With each book Peltier read, he became more curious. He soon went beyond reading and began to carry out experiments. Few scientists tackled so many different fields with such enthusiasm. He dissected animals, observed the night sky, investigated chemical processes, and forecasted the weather, among a dozen other pursuits. But the contributions for which he is best remembered today are his experiments with electricity.

In 1834, Peltier found that when he caused electrical current to flow through a circuit made from two different conductors (Figure A), something remarkable happened at the points where the dissimilar wires connect. Depending on the direction of the current flow, one junction got hot and the other grew cold. The more voltage applied, the hotter and colder things got.

This phenomenon is now known as the *Peltier effect* and is the key idea in the design of many precision instruments, satellites, heat pumps, dehumidifiers, and even wine coolers.

The scientific principles behind the Peltier effect are complex, but in a nutshell, they work like this: For a given voltage, the amount of energy that electrons possess as they travel through conducting wires is different depending on the electrical conductor material. At the electrical junctions between different types of conductors, the smooth flow of electrons is interrupted, causing the equivalent of an electron traffic jam. At one side of the jam, electrons expel excess energy to the surroundings so they can enter the new conductor. Here it gets hot. At the other junction, the electrons do the opposite: forced to absorb energy from their surroundings, they make that side cold.

It wasn't long before engineers and scientists figured out that this ultra-simple circuit (merely a couple of different wire types soldered together and a battery) had a lot of interesting uses. With only a voltage source and two types of conductors, it is possible to make an electrical device that heats and cools, with no moving parts.

In this edition of Remaking History, we'll use Monsieur Peltier's thermoelectric principle to construct a desktop drink coaster that can both heat and cool a cup — your choice — at the flip of a switch.

BUILD A PELTIER HOT/COLD DESK COASTER

Refer to the Assembly Diagram (Figure B) for all steps that follow.

1. Using the holes on the fan body as a template, mark locations for drilling holes on the flat side of the aluminum heat sink. Drill ⁵⁄₁₆" diameter holes through the body of the heat sink as shown in Figure B.

2. Bend the aluminum strips into L shapes and drill a ⁵⁄₁₆" hole through each. You will need to form the L based on your particular beverage cup. See Figure B for how they'll be placed on the top of the device.

3. Assemble the L-shaped aluminum strips, heat sink, and cooling fan into a single assembly using the ¼" bolts, as shown, using the two small adjustable wrenches or screwdriver. Place washers between the bolt head and the strips, the fan and the heat sink, and the fan and the nut, as shown in Figure C and the assembly diagram.

C

WILLIAM GURSTELLE's book series *Remaking History*, based on this magazine column, is available in the Maker Shed, makershed.com.

Adobe Stock – kungverylucky, Adobe Stock – Juulijs, Adobe Stock – Archivist, William Gurstelle

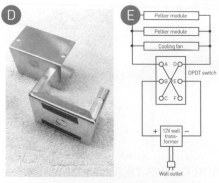

4. Drill a hole to fit the stem of the DPDT switch in the center of your project box. This is usually ½" in diameter, but measure your switch stem to make sure. Drill a similar hole in the opposite wall of the project box to run the wires as shown in Figure D.

5. Connect each Peltier module to the 12-volt wall transformer and mark which side of the unit gets hot and which gets cold.

6. Use the thermal adhesive to glue the Peltier modules one atop the other, hot side down, to the flat side of the heat sink, as shown in Figure B. The doubling of the modules provides a larger heating and cooling effect than a single module.

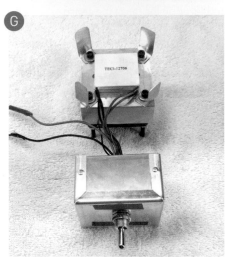

7. Use the wire cutter and stripper to wire the device as shown in the Electrical Diagram (Figure E) so that the coaster has three positions. When the switch is in the up position, the cooled sides of the Peltier units are on top. When the switch is down, the Peltier units will have warm sides up. When the switch is the middle position, the device is off.

Figure F shows how the "crossover" wiring connections are made between terminals on the DPDT switch.

Close up the project box. Your Peltier Coaster is complete (Figure G).

GO AHEAD — BE HOT AND COLD
To use your Peltier Coaster, plug the wall transformer into a power outlet and flip the switch to the up position. Hold your hand near the top surface of the Peltier module to detect whether it becomes cold. Flip the switch to the down position to check that it heats.

WARNING: The surface can become very warm or very cold quickly. Avoid touching the surface for more than an instant.

If your coaster does not heat or cool, check your connections and make sure the device is wired correctly.

Metal cups work best with the Peltier Coaster (Figure H). Enjoy your warm — or cold — beverage! ◢

MORE PELTIER-EFFECT PROJECTS

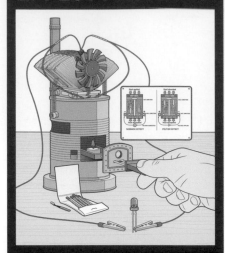

The Amazing Seebeck Generator
The Seebeck effect is Peltier's in reverse: Get 5V power from a candle flame. A classic from *Make:* Volume 15. makezine.com/go/seebeck-generator

Peltier-Cooled Cloud Chamber
Atomic punks — build a DIY cloud chamber to make radioactive particles and gamma rays visible. makezine.com/2010/05/06/how-to-build-a-peltier-cooled-cloud

Chilled Drinkibot
Peltier + pump = chill and dispense your favorite bevvie, temp-controlled by an Adafruit Trinket. makezine.com/2017/11/07/build-a-thermoelectric-cooled-drinkibot

William Gurstelle, Timmy Kucynda, Rich Olson, John Park

Upcycle a TV satellite dish and give back to nature

Written and photographed by John Pedersen

Sat-Dish Bird Bath

TIME REQUIRED:
3–4 Hours

DIFFICULTY:
Easy

COST:
$65-$75

MATERIALS
» **Satellite dish**
» **Electric submersible pump**
» **Hose to fit pump**
» **Spray paint**
» **Overflow tube** I used a snap-off extension nipple used with ground lawn sprinklers.
» **Cable tie** aka tie wrap or zip tie
» **Plastic bucket, 5gal**
» **Spray nozzle** I got one from another pump set.

TOOLS
» **Flat blade screwdriver**
» **Phillips screwdriver**
» **Pocketknife**
» **Hacksaw**
» **Electric drill with stepped drill bit**
» **Level, 36"**
» **Respirator with combination HEPA / chemical cartridges**

What to do with an old TV satellite dish, rather than just going to the scrap metal bin? I'd been thinking about it for a while: just orient it 90 degrees or so and it would make a great bird bath, and using the arm and LNB housing it could even have a shower attachment! It sits on a 5 gallon bucket as a reservoir, and the water is circulating so there shouldn't be any issues with mosquitos.

1. PREP THE DISH
Find the center of the dish, mark it, and drill a hole to fit your overflow tube (Figure Ⓐ).

2. DISASSEMBLE THE LNB
Disassemble the LNB with a blade screwdriver and pocketknife (Figure Ⓑ). Remove the feed horn and amplifier. The bracket holding the LNB was screwed together with two Phillips screws.

3. PAINT
Parts were all spray painted with about ½ a can of paint. The brand I use is Rust-Oleum Painter's Touch 2X Ultra Cover. It has stood the test of time (decades) on many of my outdoor projects.

4. MEASURE THE OVERFLOW TUBE
Put the overflow tube in the drilled hole.

Then place the level across the dish, mark the tube where the level is, and cut it off below that, so the bird bath won't overflow (Figure Ⓒ).

5. ASSEMBLY
Attach the pump adapter and nozzle to the hose. Thread the hose through the LNB and set the nozzle just a bit inset from the edge, then secure the hose in place with narrow tie wrap. Clip the LNB housing back together, then attach it back into its mounting bracket. Thread the hose down the arm and then attach the mounting bracket to the dish arm.

The hose then attaches to the pump and is put into the 5gal bucket filled with water. I had to slightly modify the bucket so that the satellite mounting bracket would fit level on it.

6. TEST IT
Fill the dish with water to just below the overflow opening. Make sure the bucket and dish assembly is on a level surface such as a deck or patio.

Using an extension cord plugged into a GFI (Ground Fault Interrupter) socket, plug the pump cord into the extension cord and enjoy your bird bath. ⊘

Ⓐ

Ⓑ

Ⓒ

bird: adobe stock - nataba

JOHN PEDERSEN
has written 44 Instructables. He lives in CehNehDeh (aka Canada).

TIME REQUIRED:
1–2 Hours

DIFFICULTY:
Easy

COST:
From $20

TOOLS
» Smartphone with internet access
» Adobe Capture CC App and free Adobe account
» Computer (optional)

Worth Repeating

Use this simple kaleidoscope app to create your own custom fabric prints from images **Written and photographed by Poppy Mosbacher**

When I couldn't find the exact fabric I wanted, I started designing my own. It was challenging to start off with a completely blank canvas, until I discovered the Adobe Capture app, which is free to use. The app works like a digital kaleidoscope, turning anything that you point your phone's camera at into a repeating pattern. On-demand printing websites will print small lengths or you can fit several designs on a bigger piece of fabric.

1. CREATE A REPEATING PATTERN FROM A PHOTO

Open Adobe Capture CC on a smartphone. Scroll through the functions near the bottom of the screen and select Patterns.

TO CREATE A NEW PHOTO:

Hold up the phone and point it at objects or scenery. The square in the center is what the camera can see. The app uses a segment of that square to create the pattern shown in the rest of the screen (Figure A). You can change the shape of the segment to create different patterns: Just click on the icon in the top right corner and choose a new shape. Then click the circle at the bottom of the screen to take the picture.

TO USE AN EXISTING PHOTO:

Click on the photo icon in the bottom right corner. Select the location where the photo is stored. Click on the photo to select it. You can then adapt the pattern by changing the segment shapes, same as creating a new photo. Then click the tick (checkmark) button at the bottom of the screen.

You can rotate the photo using the scroll wheel or the icon in the bottom left corner (Figure B). To undo, click the icon on the bottom right. Click Save.

Create more patterns by repeating these steps, or click the X icon at the bottom left corner to view your library of patterns.

2. EXPORT THE PATTERN

From the library, click on a pattern to select it, and click on the Export icon at the bottom left corner of the screen. Click Export As→ Pattern Tile (Figure C) and follow the steps on the screen to email it to yourself. You can choose other options to save it, but make sure it's as a Pattern Tile, otherwise the edges won't line up later.

Download the pattern tile from your email. I did this on my computer to have a larger screen for the next step, but it can all be done from a smartphone.

3. UPLOAD TO FABRIC PRINTING WEBSITE

Choose an on-demand digital printing website. As I'm in England I used Contrado

(contrado.co.uk) but when I'm in the U.S. I would use Spoonflower (spoonflower.com).

Choose a fabric. It can be helpful to order swatch packs so you can feel the fabrics first (Figure D). Upload the pattern tile you created earlier (Figure E).

Set the size of the fabric. In the preview area of the screen, drag the corners in to make the pattern tile a bit smaller.

In the Effects section, choose a Pattern Repeat Style and the pattern will be repeated to fill in the rest of the fabric (Figure F). If you change the size of the pattern tile, the overall pattern will update too.

Click Preview and Buy, then follow the steps to make the purchase.

TIP: Your computer screen is backlit, so your fabric might appear duller when it arrives. Print a paper copy of your pattern to get a much more realistic idea of what it would look like, and adjust accordingly.

COLOR YOUR WORLD

You can use your printed fabric to make clothes, lampshades, or cushions. To be more adventurous you could also print onto organza for wedding favors or neoprene to make your own wetsuits. ◗

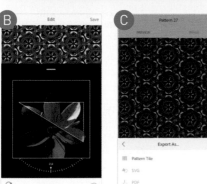

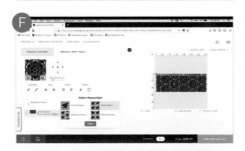

POPPY MOSBACHER
represented Britain in a robot-building contest in Japan, aged 8, after her design won a national competition. She now has a collection of robots, including a homemade ride-on elephant. A former director of Build Brighton Makerspace, she is currently helping to start a Tech and Textiles Makerspace in Devon, England.

TINKERING WITH TEXTILES

Try out these great fabric projects from the team at *Make:*, ranging from simple upcycling to stunning wearable tech.

COLORFUL SCRAP TWINE

Twist those fabric scraps and offcuts into an upcycled alternative to rope and cord, for use in weaving and other crafts. makezine.com/projects/twist-fabric-scraps-colorful-twine

3D SHIBORI TEXTURE

Make your own shape-memory fabric! Exploit a synthetic fabric's thermoplastic qualities by wrapping, folding, and binding it into shapes with thread and then boiling it. Once dried, the fabric will maintain the shape. makezine.com/go/3d-shibori-texture-fabric

FABRIC MASTERPIECE

Re-create your favorite painting as a cloth collage using your laptop, a digital projector — and the artist's permission, of course. No sewing necessary! makezine.com/projects/fabric-masterpiece

CUSTOM LED MATRIX HANDBAG

This bright, beautiful, and interactive tote displays low-resolution animations, scrolling text, and even Twitter data in real-time. Created by Debra Ansell — see her newest project on page 56 of this issue. makezine.com/projects/led-matrix-handbag

The GoonieBox

How I built my intriguing, interactive puzzle table inspired by *The Goonies*

Written and photographed by Guido Bonelli

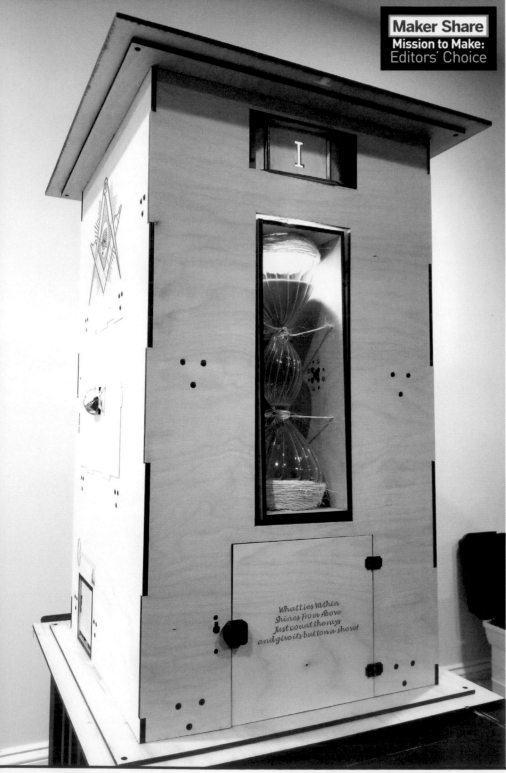

Maker Share
Mission to Make:
Editors' Choice

I wanted to create a unique piece of furniture for my home that my guests could interact with. I also wanted mysterious scriptures scribed all over the sculpture to intrigue my guests, to lure and enchant them like a moth to a flame. But, where would I come up with something like this? I needed a muse.

What inspired me was a puzzle from the first installment of the game *The Room* by Fireproof Games. If you've never played it, after reading this head immediately over to Google Play or the App Store and download all three games. They are jaw-dropping in graphics, mind-bending in puzzles, and unique in game play.

Imagine walking into a room and seeing an intricate piece of furniture ("The Safe") sitting there with unassuming elegance, adorned with buttons, gears, and secret doors hidden throughout its structure screaming to be inspected. Immediately upon playing the game I thought, "I must build one!" It wouldn't be till years later that I came up with a design worthy of this puzzle.

A COUCH, A CLOCK, AND A CHILDHOOD MOVIE

Fast-forward to December 2015 when I was looking for a new couch. I happened upon a very curious clock that stood over 6 feet tall, and was made of solid metal with a giant hourglass in the middle. As I walked around the clock, I touched the steampunk-esque metal wheel and to my surprise, it moved. Not only did it move but actually turned the massive hourglass in the middle! I began to conceive an idea that would meld my love for clocks and puzzles in one unique piece of furniture.

As the idea began to rummage through my head, I started to imagine how I could incorporate an hourglass into my design. After scouring Pinterest, Facebook, and the very bowels of the interwebs (scary stuff down there) for the perfect hourglass, I had found one (Figure A)! It was carried by a store called Anthropology. Like a SpaceX rocket, I hurled my body toward the closest brick and mortar store. Success, they had it in stock! An hourglass wouldn't be enough to tell time, though; it needed to display the hours as well. I didn't want to simply surround it with numbers and point to them somehow; the thought of making another infinitely gyrating, number-hunting arm built into my *Mona Lisa* was not appealing.

Then it hit me! It needed a Roman numeral dial with a hint of RGB backlighting, of course (Figure Ⓑ). Duh, right? I know!

As the outer box started taking shape, I realized that there was plenty of room to contain "stuff." But what kind of stuff? Well, puzzles, of course.

My eyes started darting back and forth conjuring images long since passed. In a flash of brilliant white light, an old 35mm film reel began to churn in my mind. It was from my favorite childhood movie, *The Goonies*: "Sloth love Chunk!" To this day, that movie makes my inner child want to grab the nearest ruler and shake it in the air like a pirate's sword.

With that, I decided to make my part clock, part board game, part furniture piece a treasure hunting game! Complete with lasers (Figure Ⓒ), skeleton keys, and (shhhhhh …) a treasure vault.

I bet you're wondering what the treasure is? Well, for that you would just have to play to find out.

MY DESIGN PROCESS
This can be applied to anything:
1. I start with my "muse," which for me is puzzles, clocks, nature, and meddling with electronics.
2. From there I render everything in 3D, and print out what is feasible on my MakerBot Replicator 1 (Figure Ⓓ). These 2 steps are absolutely critical in getting out all of the bugs.
3. Once I'm happy with the form and fit of my prototype (Figure Ⓔ), I'll send out all my PCBs, and my wood to get cut, and order all parts.
4. While I'm waiting for my parts to arrive, I begin to write my code, using a hardware-simulating tool I designed, called Dr.Duino (see below right).
5. Once all the parts come in, I begin the integration phase (Figure Ⓕ). Since everything has been worked out from the steps described above, the model usually comes together with very little trouble.

FLAWLESSLY FINE-TUNED
The final GoonieBox, believe it or not, didn't have one single design mistake. Now, don't get me wrong, during the 3D modeling and printing phases, I had billions of mistakes. However this process allows me to minimize that tremendously. ◗

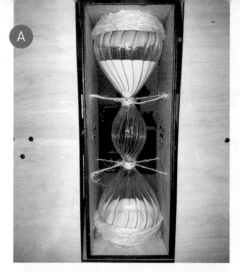

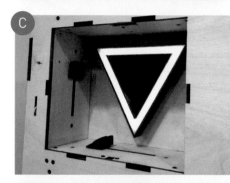

Dr.Duino (drduino.com) is a prewired Arduino learning/hardware debugging kit. GoonieBox's DNA is made up of Dr.Duino — it allowed me to write all GoonieBox's code before I had any actual hardware.

TIME REQUIRED:
800 Hours

DIFFICULTY:
Advanced

COST:
$1,500–$2,000

MATERIALS
» **ICStation Mega2560 microcontroller** Arduino compatible
» **Real-time clock (RTC) chip**
» **DC motors (2)** controlled by PWM
» **RobotGeek servos (5)** continuous rotation
» **Texas Instruments motor drivers**
» **Slide potentiometers**
» **NeoPixel RGB LEDs (60)**
» **Plywood**
» **Hourglass, skeleton key, lasers, other parts for puzzles**

TOOLS
» **Laser cutter** or a laser cutting service
» **Computer** for 3D modeling and coding
» **3D printer**
» **Dr.Duino test platform** for developing and debugging hardware and code

GUIDO BONELLI is an inventor/entrepreneur and founder of the Dr.Duino brand, focusing on creating learning/development kits for Arduino.

[+] Check out the project page at makershare.com/projects/gooniebox-part-clock-part-mysterious-boardgame and watch the clock in action at youtu.be/YK7x2fiSgZs.

Toy Inventor's Notebook

SPIN 'N SPILL BALANCING BOT GAME

Invented and illustrated by Bob Knetzger

BOB KNETZGER is a designer/inventor/musician whose award-winning toys have been featured on *The Tonight Show*, *Nightline*, and *Good Morning America*. He is the author of *Make: Fun!*, available at makershed.com and fine bookstores.

TIME REQUIRED:
30 Minutes

DIFFICULTY:
Easy (especially with a laser cutter!)

COST:
$5

MATERIALS
» Plywood or MDF board, ⅛"×8½"× 11"
» Quarters (up to 9)

TOOLS
» **Laser cutter** Get a Glowforge discount at glowforge.com/bobknetzger!

Here's a fun and easy action game: the Spin 'N Spill Balancing Bot. Stack the gears onto the tippy balancing robot — but be careful! If you cause him to drop or spill, you lose!

This game is easy to make, especially if you have access to a laser cutter. ⅛" MDF is a good material to use: light and strong (and cheap!), and easy to get the balancing right.

CUT PARTS

Look online at makezine.com/go/spin-n-spill-game for some useful files. There's an Adobe Illustrator file with vector art for the laser cutter tool path and also bitmap art for engraving the Makey bot design. If you have access to a Glowforge laser cutter and some of their Proofgrade brand medium Draftboard MDF, jump right to the *.svg* file and you're ready to go (Figure Ⓐ). If not, you may have to adjust the dimensions of the slots and tabs to match your laser cutter's kerf (the amount of material burned away).

Also, you may need to adjust for the actual thickness of your chosen material.

NO LASER CUTTER? You can still make this game. Simplify the shapes for easier cutting by hand, e.g., just cut a big circle around each gear instead of cutting all the teeth. The only dimensions that are critical are the slots and tabs for assembly. Cut the slots undersize at first, then trim or file for a snug fit.

ASSEMBLE

With all the parts cut out, assemble them by matching the lettered slots and tabs. See the assembly sketch (Figure Ⓑ) for help. At the bottom of the bot is a sliding compartment, which holds up to 9 quarters as a counter weight. Hold the coins in place while sliding in the smaller part, trapping them all together. Test your bot by balancing it over the edge of a table. You can add or remove coins to adjust the tippiness.

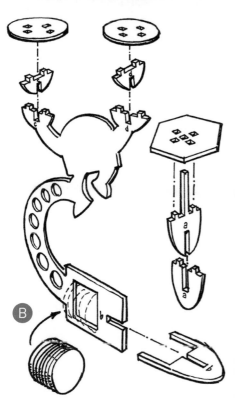

PLAY

Spin the spinner; you'll get 1, 2, 3, or SKIP. Place that number of gears onto the bot's hands without touching him or impeding his swing and sway. Be careful: If anything falls off, you're out! Placed all your gears OK? Then the next player goes. Keep spinning and stacking until somebody spills — oops, they're out! Play again until only one player remains: the winner! ◗

MORE GLOWFORGE PROJECTS

I found the Glowforge much easier to use than other laser cutters; you upload your design to the website, then the cutting plan is sent to your cutter via Wi-Fi; which lines are "cuts" or partial "engraves," each with appropriate power and speed settings. I very seldom modified the plan.

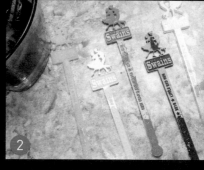

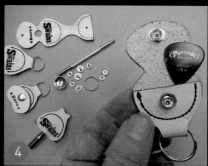

① PUMPKIN PARTY

Nonartists and technophobes can quickly create perfect custom acrylic accessories for their jack-o-lanterns. Simply draw a quick sketch on a piece of paper, and the Glowforge's built-in camera will take a picture of it and create a plan to cut out and engrave that shape. No vector art software or skills needed! This "instant result" feature is an easy entry to laser cutting for any age or skill level.

② SWIZZLE STICKS

I created this set of custom swizzle sticks for my country band, The Swains, in ⅛" acrylic. It's the perfect thickness, and the final result looks just like a real, injection molded swizzle stick. Laser-cut acrylic is professional looking, with "flame polished" glossy edges. The text on the clear version is reversed and engraved from the back, which gives a nice finished effect.

The Glowforge camera helps you place your designs onto the material with an intuitive drag-and-drop feature. You can rotate and resize, and cut-and-paste for quickly cutting out multiples.

③ CAKE TOPPERS

These are very popular to make and are perfect for personalizing a birthday cake. I was amazed to be able to cut such a delicate design. The width of the filigree and tiny leaves is only 0.010"!

④ PICK POUCHES

Leather cuts and engraves nicely to make these guitar-pick pouch key rings — each individually personalized for the whole band. The drummer's version holds a drum key.

⑤ PATENT PLAQUE

I engraved the text and illustrations from a patent into clear acrylic and then placed it over a black and white photo printed on blue paper. This gives a nice "floating" blueprint effect. The handsome ⅛" walnut plywood stand holds the panels together for display.

⑥ BOOK RACK

It's engraved with the title and characters from the artist's line of comics, and it knocks down for easy travel to the next book show.

BREVILLE PIZZAIOLO
$800 breville.com

Word has it that the creator of a popular music-making software title later expressed regret for ruining popular music by making it so easy to create. After a day using the Breville Pizzaiolo, having cranked out 16 perfectly baked pies, I wondered if this little countertop oven is the equivalent for the pizza making world.

I wanted to test this thing out because of its "hack mode," a hidden ability to override its presets and customize the cooking configurations — an option we don't see enough on modern appliances. But it works so dang well straight out of the box that this was hardly a need. Pan, New York, and Neapolitan styles came out just like from a restaurant; its electronic elements can crank up to 750°F to truly recreate a cooked-in-2-minutes wood-fired result, complete with leopard-spotted crust and bottom.

The ease of use may take some of the magic away from wrestling with a real brick oven (which you can and should build yourself: makezine.com/projects/quickly-construct-wood-fired-pizza-oven), but that's probably fine, especially because you'll still have to perfect your dough technique to get the best results possible. And for the price, you've probably earned it. —*Mike Senese*

ROBOROCK S5 VACUUM **$600** en.roborock.com

The Roborock S5 highlights the advancements in Roomba-style technology with its bevy of sensors and connectivity. Unlike early robot vacuum models that would do random patterns to hopefully eventually get full coverage of a room, the S5 uses its top-mounted lidar to create a map of your space, and then generates an overlapping sweeping pathway to vacuum all areas evenly (while leaving a pleasing, professional-looking pattern on your carpets). It's not a fast process — what takes 74 minutes in my home I could manually do in 15 — and it doesn't have quite the suction power of my Kenmore, but letting it pass through the house every 3-4 days still fills up its bin with a surprising amount of dust and debris. Meanwhile, it hasn't gotten stuck once, as it will reorient itself to its lidar map if jostled. It's seriously smart, and wonderfully easy.

One cool thing to do with the rig is integrate it into a connected home. With a little bit of coding, you can trigger the S5 with voice commands. "Siri, vacuum the living room" will surely garner high-fives from the family. There are various resources online for this; we like the open-source Home Assistant setup: home-assistant.io/components/vacuum.xiaomi_miio —*Mike Senese*

Mike Senese, Xiaomi, Mark Madeo, Kreg, Panavise

KREG JIG R3

$45 kregtool.com

The Kreg pocket hole jig is an incredibly simple tool that manages to make a huge impact. Pocket holes, where you hide a screw inside a hole in your wood, aren't a super new idea, nor are jigs to create them. However, Kreg's system just simply works.

You place the jig where you need it, clamp it on, drill with the supplied drill bit, then when you're ready to assemble you've got a pre-drilled hole at the right depth with the proper width for the head and shank of the screw you're using. It feels like cheating.

Building simple boxes, drawers, and even basic furniture is so much easier when things go together with minimal effort thanks to pocket holes. The hard-core traditional woodworker may sneer at using tools like these but they can sneer all they want while I enjoy the fruits of my labor!

I enjoyed this model enough that I'm looking at the fancier ones which mount to your workbench for my next purchase. I do like the portability of this tiny one though, as much of my work right now is not being done inside my shop. —*Caleb Kraft*

PANAVISE JR. 201

$30 panavise.com

When it comes to workholding on my electronics workbench, it is difficult to beat the Panavise 201. This thing is super mobile — I can plop it wherever and adjust it to hold my projects at the perfect angle, but thanks to the beefy base, it doesn't wobble or scoot around as I try to work on it.

The weighty and solid construction has actually led to a secondary use for this, as a camera stand. I can toss my GoPro, cellphone, and even my DSLR on this thing and aim it at exactly what I want — perfect for the maker who is sharing their projects online. I'd highly recommend this as a multipurpose workbench tool. —*Caleb Kraft*

REDBOARD EDGE **$22** sparkfun.com

The idea of the RedBoard Edge is simple: Take the classic Arduino Uno, and move everything around to be more conveniently located to build into the final version of your gadget, rather than sprawled out for prototyping.

On one side of the board are all the parts you'd leave sticking outside of a case: Four status lights in different colors, for power, debugging, and data transmission/receipt; a power plug with a dedicated on-off switch; and the reset button. Just 3D print or craft a case, leave that edge sticking out, and you're good to go.

The opposite side of the board has all the "pins," analog and digital, to control the guts of your project. This is the end that's laid out to be hidden inside your enclosure. These are bare solder rings, no headers. The Edge is made to be the final controller board for your gadget. If you want to prototype with it, you can order headers separately and solder them on yourself.

This board is great if you've got a gizmo you've already prototyped on a conventional Uno or similar Arduino-alike, and now you want to make a permanent version in a handsome housing. Just add housing, solder on the needed bits, upload the program you prototyped, and this tidy board is ready to go. —*Sam Brown*

For tons more board reviews, visit our online Makers' Guide to Boards: makezine.com/comparison/boards

SHOW & TELL

Get inspired by some of our favorite submissions to the Make: community

If you'd like to see your project in a future issue of *Make:* magazine submit your work to makershare.com/missions/mission-make!

① THE ULTIMATE BITCOIN TRACKER

If you've been craving a way to make your enthusiasm for Bitcoin more tangible, **Jonathan Pereira** built a physical LED Bitcoin dashboard using a Raspberry Pi, a 64×8 LED matrix, and laser-cut MDF for the enclosure. The rig cycles through information like current price, total Bitcoins left to mine, blocks until the reward is halved, hash rate, and more. Take a look at Pereira's code at the project page. makershare.com/projects/ultimate-bitcoin-tracker

② CNC ROUTING FRUIT!

What's the best way to use an expensive, precision machine like a CNC router? Clearly to put funny shapes in fruit! **Kent Watson** got the bright idea to use a CNC router to carve detailed images on the face of an apple. Also, it's a really messy way to make applesauce. Watson's trial and error-laden video is worth a watch. makershare.com/projects/cnc-routing-fruit

③ JUNKYARD DESK CHAIR

Shocked at the high cost of a new desk chair after the back fell off his old one, **Al MacKay** found a leather captain's seat from a minivan and attached it to the perfectly usable lower half of his old chair. The new seat was easy to find at his local auto salvage yard, much cheaper than its cubicle-bound cousins, and more comfy to boot! makershare.com/projects/junkyard-desk-chair

④ LITHOPHANE LAMPSHADE WITH 4-AXIS CNC

Lithophanes are thinly etched artworks that are lit from the back to reveal an image. **Lex Lennings** used a 4-axis rotary CNC milling machine to etch a family portrait into PVC pipe for a beautiful lamp. Take a look at the project page for a primer on designing cylindrical lithophanes for milling. makershare.com/projects/create-lithophane-lamp-shade-4-axis-cnc

[+] Read about our Editors' Choice, "The GoonieBox," on page 72.